SPACE ELEVATOR EARTH PORTS

Book 4 Space Elevator 2020 series

Linda Phillips

CONTENTS

Title Page
Space Elevator Earth Ports
1. Reflection   1
2. An airport for space: the Elevator Space Port   2
3. Satellites   22
4. Geostationary versus geosynchronous orbits   25
5. Orbital eccentricity   27
6. Crowded GEO space   31
7. Finding a GEO slot   32
8. Meteorological considerations   34
9. Other considerations   41
10. Elevator Space Port - preferred locations   52
11. Best South Pacific Locations   80
12. Other global locations   88
13. North Pacific   89
14. Indian Ocean   91
15. Southern Atlantic Ocean   95
16. Northern Atlantic Ocean   102
17. Best locations   105
Space Elevator Earth Ports   107

SPACE ELEVATOR EARTH PORTS

Book 4 in the Space Elevator 2020 series

The concept of a Space Elevator started over a century ago but started to move from science fiction to reality in the 1990s with the promising advent of a super strong material, carbon nanotubes, that theoretically had the strength to build a ribbon into space.

NASA started investigating the concept which led to the publication of the first detailed analysis of whether it was possible, written by Dr Brad Edwards, in a NIAC paper.

Early excitement of an immediate breakthrough, maybe even constructing one by 2010, gave way to the reality of producing carbon nanotubes of the required strength, outside the laboratory, in commercial quantities. As I write, in 2020, we are still awaiting this breakthrough and the project has moved back to the late 20s, maybe the 2030s or 2040s.

But it will happen, and when it does, it will replace rockets as the easier and cheaper way to leave Earth, opening up space travel the way airplanes opened up world travel.

This series examines the technical aspects and how it will be deployed.

In book four, we return to Earth to review the possible locations for the Elevator Space Port, and cover the eight ideal locations near the Equator, in the oceans. As soon as the first elevator is completed, we will want to build a second elevator, a third ... and so on, We consider the location issues. Locations have implications for managing ribbon movement, risk and safety measures, and avoiding collisions with space debris, asteroids and these are covered.

1. REFLECTION

In 1957...
no rocket had ever traveled to space.

Yet 12 years later in 1969...
astronauts were walking on the Moon!

***

In 1903, no airplane had ever flown.

Yet, in 2018 we routinely fly around the world, to London for business, to Hawaii for holidays.

***

In the last half of the 20th century, huge, expensive rockets were the only way to travel to space.

By the 2030's or 2040's, traveling to space and the Moon by space elevator will be the norm.

2. AN AIRPORT FOR SPACE: THE ELEVATOR SPACE PORT

So, where do we locate the Elevator Space Port, our new airport for traveling into space, for the first and subsequent elevators?

It is not as easy as just co-opting Cape Canaveral. There are factors affecting an elevator location that are different to factors that come into play for rocket launch sites. As a result, we will position the space port well away from existing space ports. It's not just a busy port like an airport. It also has to act as an anchor point for the space elevator cable, for this is where it is attached to our planet.

You might say: "Locate it near me! So, I can ride it, or get a job working on it. Only not too close to where I live – I don't want to be inconvenienced by it."

In fact, that response encapsulates the paradox facing any modern project, be it an airport, a mine, or, well, a space elevator. We all use airports, but we don't want planes taking off over our head. We all use metals and minerals, but we don't want a mine as the view from our window. It's the NIMBY "not-in-my-back-yard" factor and the space elevator will be no exception.

However, we have some good news on that score: the best location is well out to sea, in a location that won't even be visible from the mainland, So, you can rest assured that we are not about to reduce property values in your area.

Of course, the downside to that is that it won't be located conveniently on your doorstep either, but let's face it, if you are going into space on a trip that is a week each way, you can take a connecting ship or flight in your stride!

A key thing to consider is the Elevator Space Port will be in use 24/7. The ribbon stays permanently in place, with, on average, one departing cable car, and one arriving cable car, each day, So, it has to be set up for continuous operation, just like an airport.

With daily departures, carrying around 20 tons each, we can shift far more goods into space than ever before. Currently, worldwide, there are about two rocket launches a week. In 2018, there were a total of 106 rocket launches.

The first space elevator will have triple the cargo carrying capacity, as compared to all rockets. Once two space elevator ribbons are in place, our lifting capacity into space will increase sixfold. By the time a possible eight space elevator ribbons are in place, lifting capacity is increased twenty-four-fold by comparison with rocketry today.

This will really put space "on the map", as it will provide the capacity to manage all current space projects, and a dozen more besides, including Moon and Mars bases, and the construction and launch of interplanetary spaceships.

At the Earth end, all of this activity will funnel through the Elevator Space Ports and these will become some of the busiest places on Earth, akin to floating airports, with attached industrial premises, plus the equivalent of the Port of New York and New Jersey adjacent, taking deliveries from all over the world. Through in perhaps a dozen cruise ships serving as hotels and recreation areas and you start to apprehend the scale. The Elevator Space Port will be like a city unto itself.

So, the location of the Elevator Space Port is a critical matter. While the attention of the previous books in this series has been focused on construction in space, this book looks at the Earth end of business and how we will manage the ribbon systems from Earth.

3. Choosing a location on Earth

In setting the location, we must first dispel a few myths, and then rule out much of the globe. The key things are:

* The Elevator Space Port doesn't have to be on the equator;
* It doesn't have to be located on land, probably on top of a high mountain.

The land based elevator on the Equator were the early assumptions of science fiction writers, including Sir Arthur C. Clarke in his seminal sci-fi book "The Fountains of Paradise" back in 1979. But today we view the most desirable location as being on a floating platform on the ocean, on or near the Equator.

The Equator is a good spot, but in practice a range of latitudes are acceptable, probably up to 30 degrees each way. By comparison, the Tropics of Cancer and Capricorn are at 23.5 degrees, and So, we have plenty of latitude in our choice of location. Having said that, the closer to the Equator the better. When the cable first descends from space, it will descend to a point on the Equator or close, depending on the effect of Coriolis forces on the free-floating cable. But once the end of the cable has been captured, it can be dragged to any preferred location.

For reasons expanded on below, the Elevator Space Port will be a floating platform on the ocean. It will not be land based at all. so, besides being reasonably close to the Equator, what are the criteria for making the location decision?

In choosing a location, these are the desirable criteria. If they all have to be satisfied, then there are only a few suitable locations on the planet. Relaxing the criteria could open up the range, but we will need to carry out critical analysis of locations and criteria before confirming or excluding any particular locations.

This is a summary of the constraining influences:

* Latitude
* Cable movement and available ocean
* Lightning
* Tropical Storms
* Shipping lanes and flight routes
* Military protection
* Safety and recovery zones
* International airport locations
* Servicing and staffing
* Environmental issues

After taking all of these into account, there are only a handful of locations left that are ideal for an Elevator Space Port. Following, these are considered one by one.

4. How near the Equator?

Latitude is the distance from the Equator, either north or south. Recall that the Equator is 0° and the poles 90°. The tropics are those locations that are +/- 23.5° of the Equator.

The base case is for the Elevator Space Port to be exactly on the Equator. After all, if a long line is spinning around the Earth, in the same equatorial plane, then the forces of gravity and angular momentum acting on it will pull it towards the Equator.

Sir Arthur C. Clarke, in The Fountains of Paradise described it thus, in describing his mythical island "Taprobani" as the ideal location: "The earth end of the system has to be at the Equator; otherwise it can't be vertical. It would be like that tower they used to have in Pisa, before it fell over."

Sure, the Equator is a desirable location for gravitational reasons. A considerable portion of the Equator is ocean rather than land, and much of it does not suffer from tropical storms or hurricanes, which are positives.

However, there are negatives too. Much of the equatorial region is subject to intense lightning, but the other big negative is above the Earth, in space, or at LEO (low earth orbit) to be exact. All LEO satellites, and the ISS, have orbits that cross the line of the Equator.

We are planning to put a solid cable into space, in a line that cuts through all LEO orbits. Our cable will need to avoid collisions with all of those orbiting things and the fewer we need to avoid the better.

That problem lessens as we move away from the Equator, even beyond the latitude of Cape Canaveral in Florida, at 28.23o N. Rockets launched from there will put satellites into an orbit that extends 28.23o north and south of the Equator (unless they are deliberately realigned).

Other launch locations have the same effect. The base for the Ariane rocket in French Guiana, South America, is at latitude 5o north, while the Russian Baikonur Cosmodrome is a northerly 45.9 degrees. So, we are never going to get completely free of orbiting satellites, but the problem fades considerably with distance from the Equator.

If the cable is located directly on the equatorial line it will point vertically into space. Away from the Equator, it will initially be at an angle to the Earths' surface, tending towards the equatorial plane but never quite reaching it. What angle? The formula is 90 degrees less latitude. (E.g. at 28 degrees the angle will be roughly 62 degrees, slanting towards the Equator.)

Unfortunately, we cannot place our anchor directly on either pole! At the South Pole, for example, our hypothetical cable would actually be lying flat, dragging along the ground, for a few thousand km.

(That's a pity. A terminal at a pole, on Earth, the Moon or any planet, could theoretically rotate at any speed we liked, since we would not be limited to matching a geostationary orbit. If only we could stop the cable from being dragged along the ground, we'd have a workable solution. However, except for the smallest of moons, the pull of gravity would be dragging the cable along the ground for hundreds or thousands of kilometers, which would be impractical!)

For practical purposes, the Space Port should be no more than about 30 degrees latitude, either north or south.

At that latitude our "launch" capacity is only 50% of the equatorial location. If this were the only factor to consider, we would look to locate it as close to the Equator as practical, but there is more to it. So, we can actually get out of the tropics altogether, if we wish, but high latitudes, such as New York or London, are ruled out. Shanghai, Hong Kong, most of east Asia, and Hawaii, are within this target area.

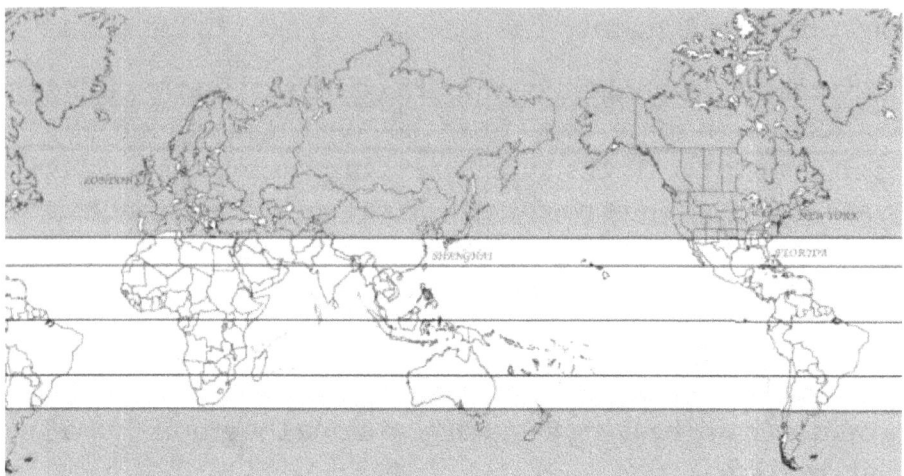

Areas greyed out (above and below the 35 degree lines) are not suitable for Elevator Space Port locations, ruled out by being too distant from the Equator.

Recall that the cable is hanging down from space – the Earth does not support it. When first deployed, the cable will be reeled out in space and dropped downward to the Earth, where it will have to be caught and held. The natural line of the cable will tend towards the Equator, because it is being dropped from GEO, and that is where it will initially be lowered. So, even if we want to deploy it at different latitudes, first we have to catch it at the Equator and then tow it north or south to the desired location.

That leads us to our second constraint: lateral cable movement on Earth.

5. Lateral ribbon movement

The cable is not rigid – it will move. A 100,000 km cable will vibrate and oscillate. The Earth precesses through its annual journey around the Sun. The Moon varies in distance from the Earth and also precesses, making small subtle changes in the gravitational field in which our cable is suspended. With all of these variations going on, the cable will exist within a normal distribution curve around an equilibrium point.

Cable cars, of varying weight and mass, will be traveling up and down the line. The weight of an object is defined as the force of gravity on the object and may be calculated as the mass times the acceleration of gravity, that is, $w = mg$.

Although weight is a function of mass, both need consideration. Wherever the car is on the cable, it will have mass, but this side of GEO, and near Earth in particular, it will exhibit weight due to the pull of gravity. Weight changes as it leaves our gravity field, but mass remains the same, excepting any mass jettisoned or used up as fuel.

They start the journey with a certain angular momentum but that will have to change as they climb. Each time a satellite is launched or captured, or a rocket leaves or joins the system, total angular momentum will be altered.

Computer controlled load management will be used to offset the varying angular momentum imposed by each car on the cable, but as it adjusts to the continual strains imposed by each movement, the Elevator Space Port will feel the pull in various lateral directions.

There will be reasons why we will purposefully move the Elevator Space Port around.

As noted above, when a cable is first deployed, it will be at the Equator by default. We will want to tow it to a pre-determined position, either elsewhere on the Equator or at different latitudes.

Also, there will be times when it needs significant movement to avoid collision with satellites or other objects in LEO, and this would be accomplished by sailing the Elevator Space Port north or south by some kilometers.

Lastly, conditions on Earth may necessitate transplanting it from one location to another in the event of conflict or other events. So, the bottom line is: the cable needs to move. This could be arranged on land to an extent. We could imagine a system of railway lines or roads that allow movement, if only there was a suitable stretch of land. But an ocean location offers practical advantages.

6. Avoiding collisions in NEO

Significant day-to-day activity will be involved in avoiding NEO (near Earth object) collisions. This is already routine for the ISS and most other satellites in LEO. Objects crossing the ISS orbit within 32 km (the safety margin) will cause the ISS to adjust orbit to avoid a collision.

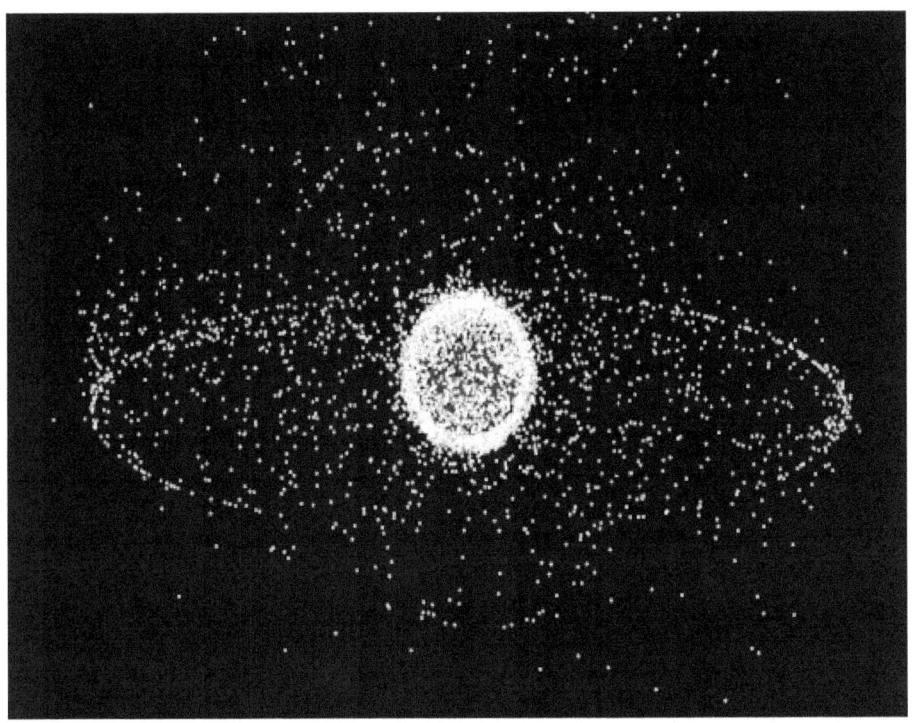

Most space objects are near Earth & at GEO, though they are spread in all directions. Image source: NASA

With the increasing amount of space junk out there, collision avoidance management has evolved into a full time activity. It isn't a unique problem for the cable, just one shared with anything else put into space. The scale of the problem increases with proximity to Earth.

Near Earth objects include the Iridium satellites in polar orbits. Iridium reports: 'On July 25, 2018, Iridium successfully launched 10 more Iridium NEXT satellites into orbit from Vandenberg Air Force Base in California. Shortly after deployment, Iridium confirmed successful communication with all 10 new satellites, formally bringing the total number of Iridium NEXT satellites in orbit to 65. This leaves just one more launch to complete this ambitious launch program.

The Iridium constellation is comprised of six polar orbiting planes, each containing 11 operational cross-linked satellites, for a total of 66 satellites in the active constellation. This unique architecture creates a web of coverage around the earth, enabling Iridium to provide real-time communications over the oceans and from even the most remote locations. One more Iridium NEXT launch is scheduled for 2018, which will bring Iridium's total number of new satellites in orbit to 75, including nine spares.'

More are to come, from other players. SpaceX recently trademarked the name Starlink for its planned network of more than 11,000 communications satellites.

In November 2018, SpaceX received US approval to deploy 7,518 broadband satellites, in addition to the 4,425 satellites that were approved earlier. SpaceX's initial 4,425 satellites are expected to orbit at altitudes of 1,110 km to 1,325 km, well above the ISS.

The new approval is for the proposal to add a very-low Earth orbit (VLEO - their nomemclature) NGSO [non-geostationary satellite orbit] constellation, consisting of 7,518 satellites operating at altitudes from 335 km to 346 km.

At the same time, the FCC approved US market entry for smaller satellite systems being built by Kepler Communications, Telesat Canada, and LeoSat. These systems consist of 140 satellites for Kepler, 117 satellites for Telesat, and 78 satellites for LeoSat.

'Development began in 2015, and prototype test-flight satellites were successfully launched on 22 February 2018. Initial operation of the constellation could begin as early as 2019 to 2020.' (Wikipedia)

How space objects are tracked

'In the USA, the Department of Defense maintains a satellite catalog on objects in Earth orbit. The United States Space Surveillance Network (SPACETRACK) detects, tracks, catalogs and identifies artificial objects orbiting Earth, e.g. active/inactive satellites, spent rocket bodies, or fragmentation debris. The system is the responsibility of the Joint Functional Component Command for Space, part of the United States Strategic Command (USSTRATCOM). About 7% of the space objects are orbiting satellites.' (Wikipedia)

NASA and the DoD cooperate and share responsibilities for characterizing the satellite (including orbital debris) environment. DoD's Space Surveillance Network tracks discrete objects as small as 5 centimeters in diameter in low Earth orbit and as small as 1 meter in geosynchronous orbit. Currently, about 15,000 officially cataloged objects are still in orbit. The total number of tracked objects exceeds 21,000. Using special ground-based sensors and inspections of returned satellite surfaces, NASA statistically determines the extent of the population for objects less than 10 centimeters in diameter.

Satview provides live satellite tracking data, including the ISS, and known re-entries.

NASA's CNEOS, the Center for Near-Earth Object Studies, computes asteroid and comet orbits from its base at JPL. Their web page reports:

'With over 90% of the near-Earth objects larger than one kilometer already discovered, the NEO Program is now focusing on finding 90% of the NEO population larger than 140 meters. Many of the charts and tables depend on diameters that can only be roughly inferred from an asteroid's estimated absolute magnitude (H) and an assumed reflectivity, or albedo.'

'The JPL Center for NEO Studies (CNEOS) computes high-precision orbits for Near-Earth Objects (NEOs) in support of NASA's Planetary Defense Coordination Office. These orbit solutions are used to predict NEO close approaches to Earth and produce comprehensive assessments of NEO impact probabilities over the next century. Continually updated calculations of orbital parameters, close approaches, impact risks, discovery statistics, and mission designs to possibly human-accessible asteroids are made available on this website and to user scripts through an Application Program Interface (API). CNEOS supports observers through the JPL Horizons high precision ephemeris computation capability.'

'CNEOS is the home of JPL's Sentry impact monitoring system, which performs long-term analyses of possible future orbits of hazardous asteroids, searching for impact possibilities over the next century. Similarly, the CNEOS Scout system monitors the MPC webpages of new potential asteroid discoveries and computes the possible range of future motions even before these objects have been confirmed as discoveries.'

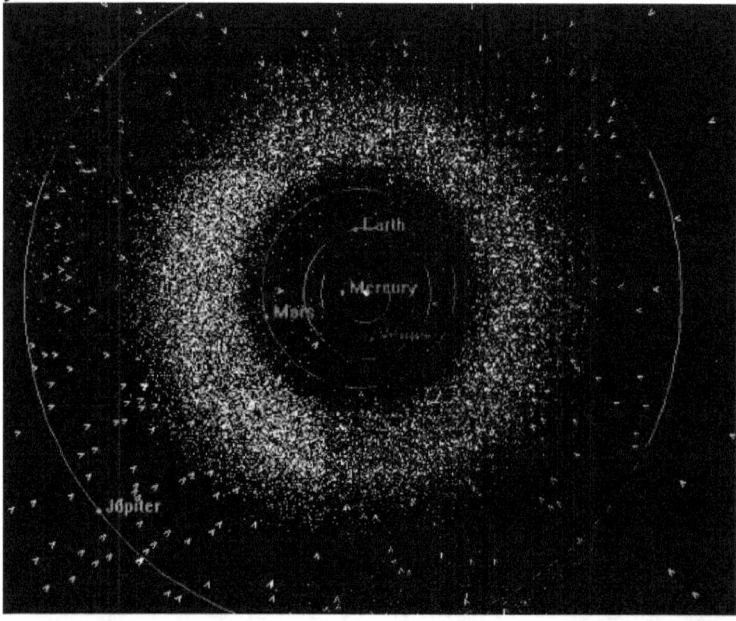

CNEOS inner solar system orbits
Courtesy NASA/JPL-Caltech

It's a reminder, while we care about the objects orbiting above Earth, Jupiter is the most massive planet in our solar system, containing more mass than all the other planets and objects put together. To a first approximation, our solar system consists of two objects: the Sun, and Jupiter. If Jupiter had been a bit bigger, it would have been a star instead of a gas giant, and this would be a binary star system. All the other bits and pieces inside of Jupiters orbit, including us, are just bits of dust by comparison, of importance only to us tiny beings.

space-track.org tracks NEO analyst objects, providing access to U.S. Government space situational awareness information.

'Analyst objects are on-orbit objects that are tracked by the U.S. Space Surveillance Network (SSN) with insufficient fidelity for publication in the public satellite catalog (SATCAT). The lack of fidelity may be due to infrequent tracking, cross-tagging (observation association with closely-spaced objects), or inability to associate the object with a known launch. Today there are approximately 17,000 on-orbit objects in the public SATCAT and approximately 6,000 on-orbit analyst objects for a total of 23,000.

The 18th Space Control Squadron (18 SPCS) is responsible for analyzing tracking data from the SSN to maintain the SATCAT, and creating and updating analyst objects as new satellites launch and new objects are found. The analyst range, which is denoted by a satellite number from 80,000-89,999, is used like an analytical sandbox, where Orbital Analysts (OA) can create, change, and update objects until they have sufficient data and information to transition them to the public SATCAT. Consequently, analyst numbers can be constantly reused for different objects.'

The 18th Space Control Squadron is based at Vandenberg Air Force Base, CA, 250 km northwest of Los Angeles, CA. The squadron is a geographically separated unit of the 21st Space Wing, Peterson Air Force Base, CO. The mission statement reads:

'MISSION
- Deliver foundational Space Situational Awareness to assure global freedom of action in space

The squadron is the newest addition to the 21st Space Wing. It is tasked with providing 24/7 support to the space sensor network (SSN), maintaining the space catalog and managing United States Strategic Command's (USSTRATCOM) space situational awareness (SSA) sharing program to United States, foreign government, and commercial entities. The squadron conducts advanced analysis, sensor optimization, conjunction assessment, human spaceflight support, reentry/break-up assessment, and launch analysis.'

These are not the only bodies tasked with tracking satellites, objects and debris in space, but they demonstrate a high situational awareness of space objects, delivering a degree of confidence we can manage the collision risk for the space elevator cable. The most significant risk comes from NEO's within 400 km of Earth.

The Space Fence

A new tracking methodology has been implemented by the USA at its missile testing base, Kwajalein, Marshall Islands, in the Pacific Ocean. In 2018, Lockheed Martin has completed integration of the U.S. Air Force Space Fence to begin tracking objects, at least in a testing mode.

According to a report in Space News:

'In early 2019, the Kwajalein Space Fence is scheduled to begin initial operations. The Space Fence will sends out a curtain of radio frequency energy wider than the continental United States.

As satellites and debris pass through the curtain, the system will detect them and determine whether the objects are already in the Space Surveillance Network's catalog.

"If they correlate with what's in the catalog, we drop them pretty fast," said Schafhauser. "But if it's something new, we track it through the entire field of regard and get a very accurate orbital determination."

In addition to keeping tabs on low Earth orbit, the Space Fence is designed to create smaller "micro-fences" in every orbit up to geostationary orbit."

As companies prepare to send hundreds or thousands of satellites into communications constellations into LEO, government agencies and commercial satellite operators are calling for enhanced space situational awareness and space traffic management.'

In addition to the base at Kwajalein, a second Space Fence is under development at Exmouth in Western Australia. The Space News report continues:

"Having a second space fence will give you more opportunities to track and ultimately understand where things are So you can help prevent collisions," said Matthew Hughes, business development manager for Lockheed Martin space surveillance programs.

Writing about the Exmouth site, the **Australian Defense Force** says:

'The Royal Australian Air Force (RAAF) will be looking a long way past aircraft ceiling height and into space by 2016, thanks to the recent acquisition of a C-Band Space Surveillance Radar.

Brought to Australia in June with the help of the US Air Force and a 36SQN C-17 Globemaster II, the radar is being installed at the remote Naval Communication Station Harold E Holt, north of Exmouth, Western Australia. Planned and delivered by the DMO through project AIR3029 Phase 1, this is the first time that the RAAF has operated such a capability.

The radar will assist the development of a Space Situational Awareness capability in Australia and will strengthen the US global Space Surveillance Network's ability to track space assets and debris.

"It will contribute to the global public good by making this information publicly available and providing satellite operators around the world with warnings of possible collisions between space objects, thereby reducing the danger posed by space debris."

C-Band Space Surveillance Radar being installed at Exmouth

7. Satellites in geostationary and geosynchronous orbits

By the time we get out to GEO, 35,500 km away from Earth, conditions are different.

Wikipedia provides a list of all satellites in geosynchronous orbit (as at 2015, updated by the UCS Satellite Database 2018) and adds the following useful information:

'These satellites are commonly used for communication purposes, such as radio and television networks, back-haul, and direct broadcast. Traditional global navigation systems do not use geosynchronous satellites, but some SBAS navigation satellites do. A number of weather satellites are present in geosynchronous orbits. Not included are several more classified military geosynchronous satellites, such as PAN.'

'Listings are from west to east (decreasing longitude in the Western Hemisphere and increasing longitude in the Eastern Hemisphere) by orbital position, starting and ending with the International Date Line.'

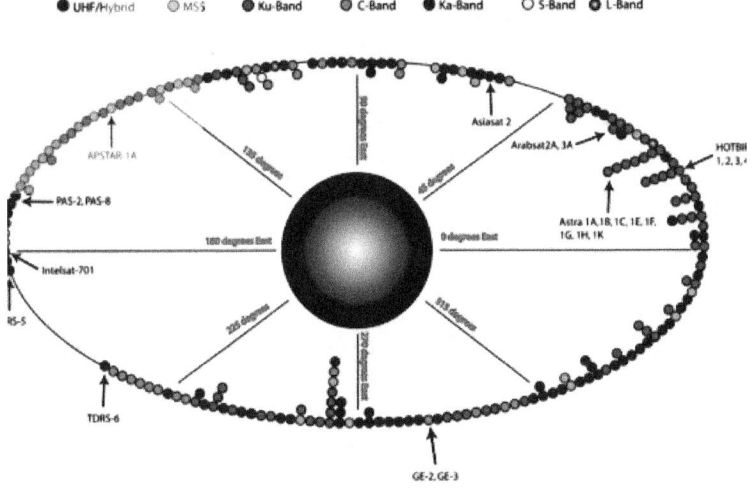

Satellites at GEO
Credit: Smithsonian National Air and Space Museum

'A special case of geosynchronous orbit is the geostationary orbit, which is a circular geosynchronous orbit at zero inclination (that is, directly above the Equator). A satellite in a geostationary orbit appears stationary, always at the same point in the sky, to ground observers.

Popularly or loosely, the term "geosynchronous" may be used to mean geostationary. Specifically, geosynchronous Earth orbit (GEO) may be a synonym for geosynchronous equatorial orbit, or geostationary Earth orbit. Communications satellites are often given geostationary orbits, or close to geostationary, So that the satellite antennas that communicate with them do not have to move, but can be pointed permanently at the fixed location in the sky where the satellite appears.'

'Some of these satellites are separated from each other by as little as one tenth of a degree longitude. This corresponds to an inter-satellite spacing of approximately 73 km. The major consideration for spacing of geostationary satellites is the beam-width at-orbit of uplink transmitters, which is primarily a factor of the size and stability of the uplink dish, as well as what frequencies the satellite's transponders receive; satellites with discontiguous frequency allocations can be much closer together.'

Since we will be very interested in satellites at GEO, to be serviced from the Home Station of the space elevator, at GEO, we can expand on the above:

'Listings are from west to east' ... The satellites are in a circle around the Earth. By agreed convention, when providing a list of GEO satellites, we start with the position above the IDT (International Date Line), known as the Prime Meridian, longitude 0o, above Greenwich, London, England.

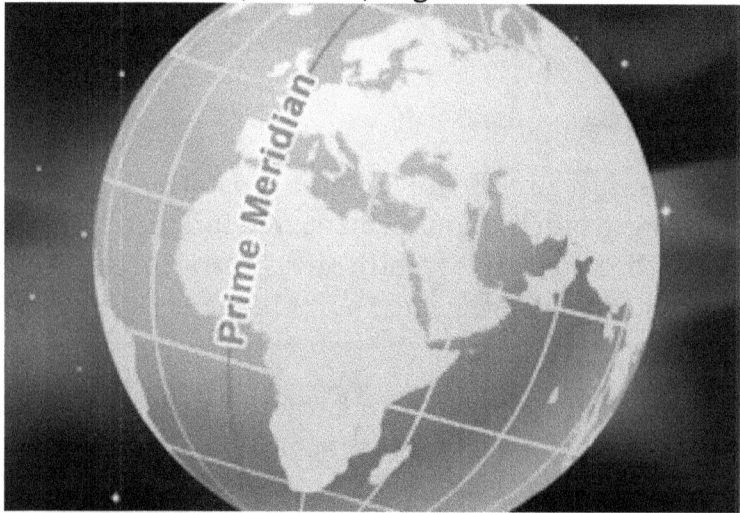

PRIME MERIDIAN. IMAGE SOURCE NASA

Visitors to London include a visit to Greenwich as a must-see, where they can stand, one leg in the West, another leg in the East.

3. SATELLITES

Satellites are then listed in order, moving west, towards the USA, onwards over the Pacific Ocean, Asia, the Middle East and finally, back to Europe and London. The UCS Satellite Database lists 1,886 known satellites, updated to April 2018. This includes all orbits and everything down to the little CubeSats, not just satellites in GEO. Of these, 548 are listed in the database as being at GEO.

Satellite Quick Facts *(includes launches through 4/30/18)*

Total number of operating satellites: 1,886

United States: 859   Russia: 146   China: 250   Other: 631

LEO: 1,186   MEO: 112   Elliptical: 40   GEO: 548

Total number of US satellites: 859

Civil: 20   Commercial: 495   Government: 178   Military: 166

Source: UCS Satellite Database

In theory, at geosynchronous orbit, the "ring" around Earth can accommodate a number of satellites — 1,800 altogether, 145 km apart, according to one analysis by Lawrence Roberts, published in the Berkeley Technology Law Review.

The circumference of the circle at GEO is approximately 260,000 km. There are 360o in a circle, So each single degree marks out an arc of some 720 km. As Wikipedia remarks, some satellites have a separation of 1/10 degree or about 72/73 km. But they are not evenly distributed. They reflect earth-based demand, So a large proportion service the USA while there is a large gap corresponding to the middle of the Pacific Ocean.

This gap over the Pacific Ocean is of interest to us, since a number of proposed Elevator Space Port locations are in the Pacific Ocean. When the Home Station is constructed at GEO, it requires sufficient clearance from other GEO satellites. How much clearance we haven't defined yet, but allowing movement of the Home Station through oscillation and deliberate repositioning, a 1,000 km or more clearance is desirable, perhaps a figure up to 5,000 km.

The Commercial Space Operations Center (ComSpOC) is a commercial satellite tracking center that is a source of tracking information. It is part of AGI (Analytical Graphics Inc.)
It provides a useful link to a Satellite Viewer.

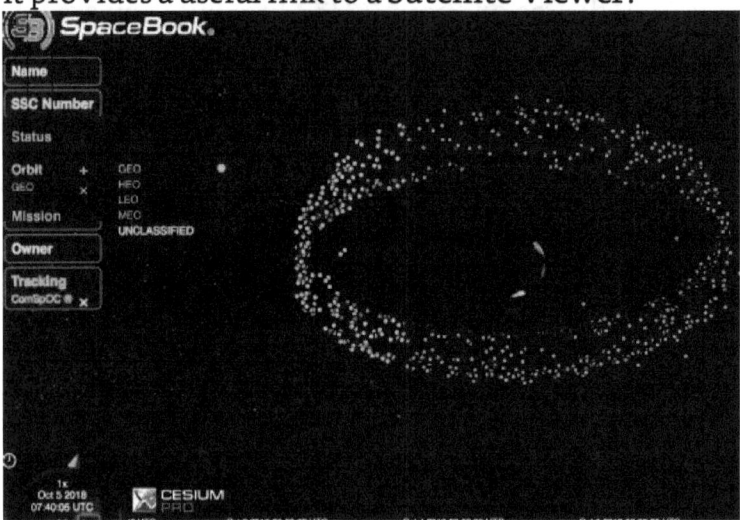

SpaceBook satellite viewer, showing satellites in GEO orbit and their inclination

Using the Satellite Viewer, we can display satellites at GEO. As can be seen, they are not all exactly on the projected line of the equator, but are in orbit in planes near to, but above and below the equatorial line, which makes for further complications when managing the Home Station which will be placed on the equatorial plane.

4. GEOSTATIONARY VERSUS GEOSYNCHRONOUS ORBITS

As explained by Dr TS Kelso,,at Celestrak, a geosynchronous orbit is any orbit having a period equal to the earth's rotational period. But this requirement is not sufficient to ensure a fixed position relative to the earth. While all geostationary orbits must be geosynchronous, not all geosynchronous orbits are geostationary. Unfortunately, these terms are often used interchangeably.

A geostationary orbit is one where a satellite in that orbit will appear to hover stationary over a point on the earth's surface. Unlike all other classes of orbits, however, where there can be a family of orbits, there is only one geostationary orbit.

First, to clarify what is meant by "the earth's rotational period." For most timekeeping, we consider the earth's rotation to be measured relative to the sun's (mean) position. However, since the sun moves relative to the stars (inertial space) as a result of the earth's orbit, one mean solar day is not one sidereal day. A geosynchronous satellite completes one orbit around the Earth in the same time that it takes the Earth to make one rotation in inertial (or fixed) space. This time period is known as one sidereal day and is equivalent to 23h56m04s of mean solar time.

To ensure a satellite remains over a particular point on the earth's surface, the orbit must also be circular and have zero inclination. The image shows the difference between a geostationary orbit (GSO) and a geosynchronous orbit (GYO) with an inclination of 20 degrees.

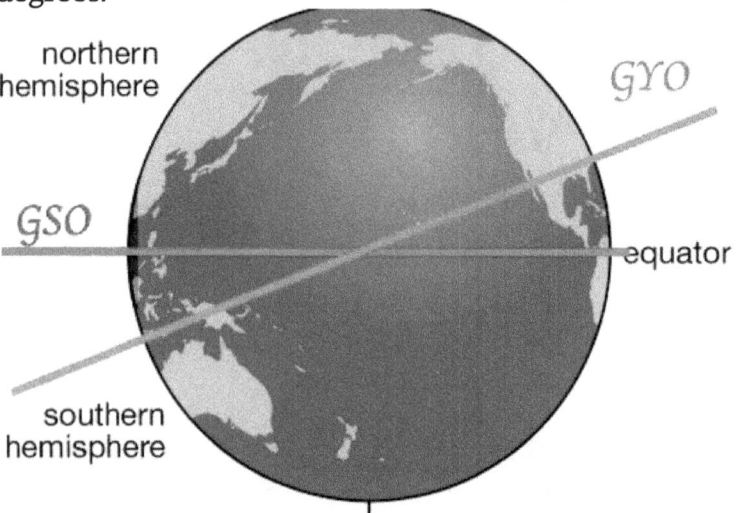

Geostationary orbit (GSO) and Geosynchronous orbit (GYO)

Both are circular orbits. While each satellite will complete its orbit in the same time it takes the Earth to rotate once, the geosynchronous satellite (GYO) will move north and south of the equator during its orbit while the geostationary satellite will not.

5. ORBITAL ECCENTRICITY

The next graphic shows the implications of each orbit. The figure-eight red ground track is that of the geosynchronous orbit (GYO). The geostationary satellite (GSO) sits fixed at the crossover point of the figure eight (over the equator). If the geosynchronous satellite has, say, an eccentricity of 0.10, the slanted teardrop shape results.

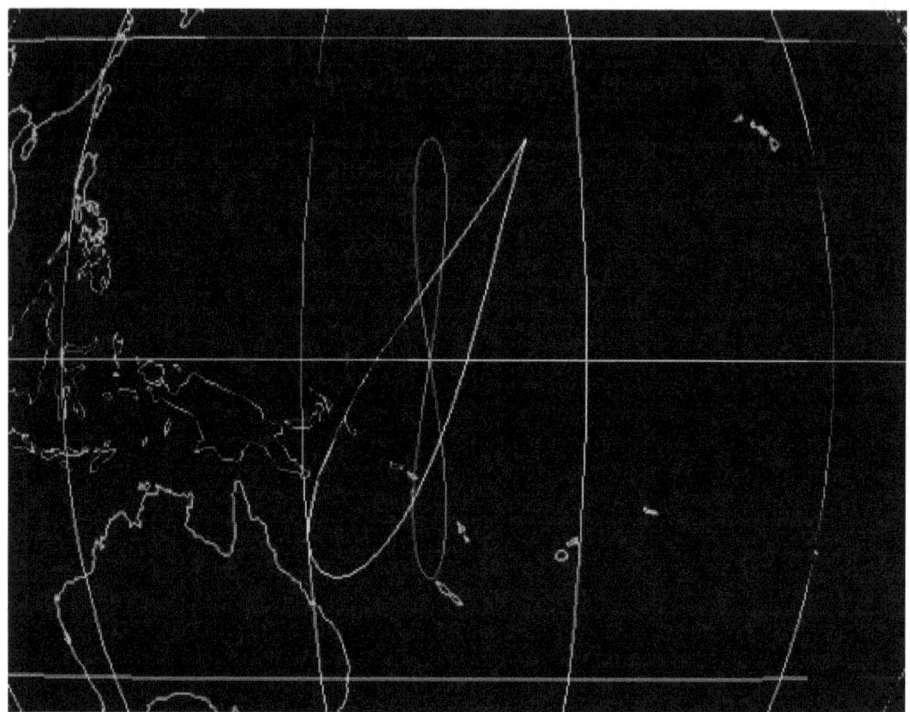

Source: Celestrak

So, a geostationary (GEO) orbit is defined as an orbit with a period equal to the earth's rotational period and with zero eccentricity and inclination. By this definition, there is only one geostationary orbit, circling the Equator at a mean altitude of 35,786 km. In this book, we often refer to GEO as being 35,500 km away, in round numbers, but 35,786 km is the exact number.

To be in geostationary orbit, a satellite must be over a point projected from the Equator. This limitation is not serious since most of the Earth's surface is visible from geostationary orbit at any one time. In fact, a single geostationary satellite can see 42% of the earth's surface at any one time, reaching latitudes between 81° S and 81° N.

Satellite stationkeeping, libration, perturbations and gyrations in GEO

New communications satellites are placed in a true geostationary orbit initially, but there are several forces which act to alter their orbits over time. The geostationary orbital plane is not coincident with the plane of the Earths orbit (the ecliptic) nor the Moons orbit. Add in the gravitational attraction of the Sun and the Moon; all these forces act to pull the geostationary satellites out of their equatorial orbit, gradually increasing each satellite's orbital inclination. In addition, the noncircular shape of the Earths surface causes these satellites to be slowly drawn to one of two stable equilibrium points along the Equator, resulting in an east-west libration (drifting back and forth) about these points.

As further explained by **SatSig**, once a satellite has been placed into its geostationary orbit position, it gradually starts to drift north-south on a daily basis due to the influence of the Sun and Moon. There is a gradual increase in the inclination of the orbit. If left alone, a satellite that has initial zero inclination will have its inclination increase at the rate of 0.8° per year, whereas an acceptable satellite station keeping box is +/- 0.15°. So, most GEO satellites need positional corrections every couple of years.

A small proportion of the total fuel is used for east-west orbit adjustments, since there is a tendency for satellite to very slowly drift sideways. Gravity is slightly stronger at three points around the Equator, being the South American Andes, East Asia and PNG, We have gravity variations to contend with.

To counteract these perturbations, sufficient fuel is loaded into all geostationary satellites to periodically correct any changes, called station keeping. North-south station keeping corrects the slowly increasing inclination back to zero and east-west station keeping keeps the satellite at its assigned position within the geostationary belt.

These maneuvers are planned to maintain the geostationary satellite within a small distance of its ideal location (both north-south and east-west). This tolerance is normally designed to ensure the satellite remains within the ground antenna beam width without tracking.

Once a satellite has expended its fuel, its inclination will increase and it will drift in longitude, a threat to other geostationary satellites. Sometimes, geostationary satellites are boosted into a slightly higher orbit at the end of their planned lifetime to prevent them causing havoc with other geostationary satellites. This final maneuver assumes that no unplanned failure has occurred, such as a power or communications failure, which would prevent management.

An advantage of having the space elevator in place lies in the ability to refuel satellites, extending their useful life. Fuel can be lifted from Earth and stored at the Home Station at GEO. From there, a refueling satellite can drift slowly around the entire GEO, stopping to refuel satellites as needed, as a sort of AAA callout in space!

6. CROWDED GEO SPACE

Returning to the locations of GEO satellites, as might be expected, most satellites are positioned to service the USA, Europe and Asia and slots covering a 300° arc are crowded. However, a significant arc over the Pacific Ocean is almost clear of satellites, which is fortunate for us, as several of the ideal space elevator slots are here.

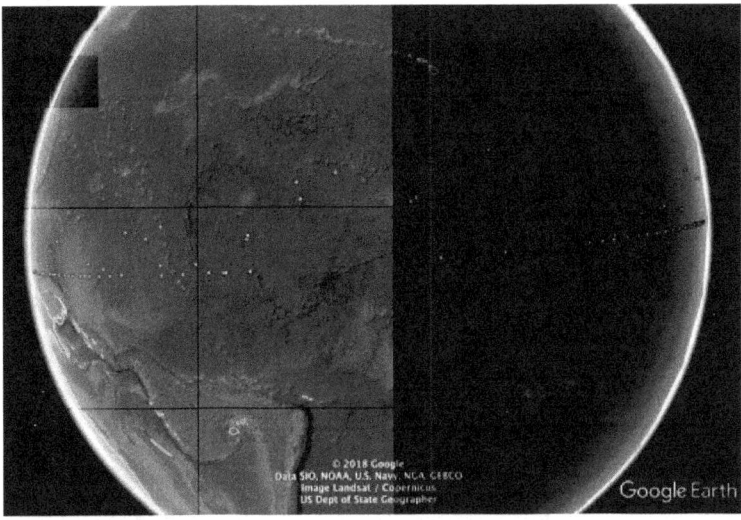

GEO satellite (yellow dots) locations projected to Earth surface
Source:
UCS satellite database as of April 2018 projected onto Google Earth

7. FINDING A GEO SLOT

That's not to say we can't position a space elevator elsewhere on Earth, but some juggling, and negotiations over slot ownership, is likely to be required elsewhere. The comparison of slots taken over the rest of the Earth shows this.

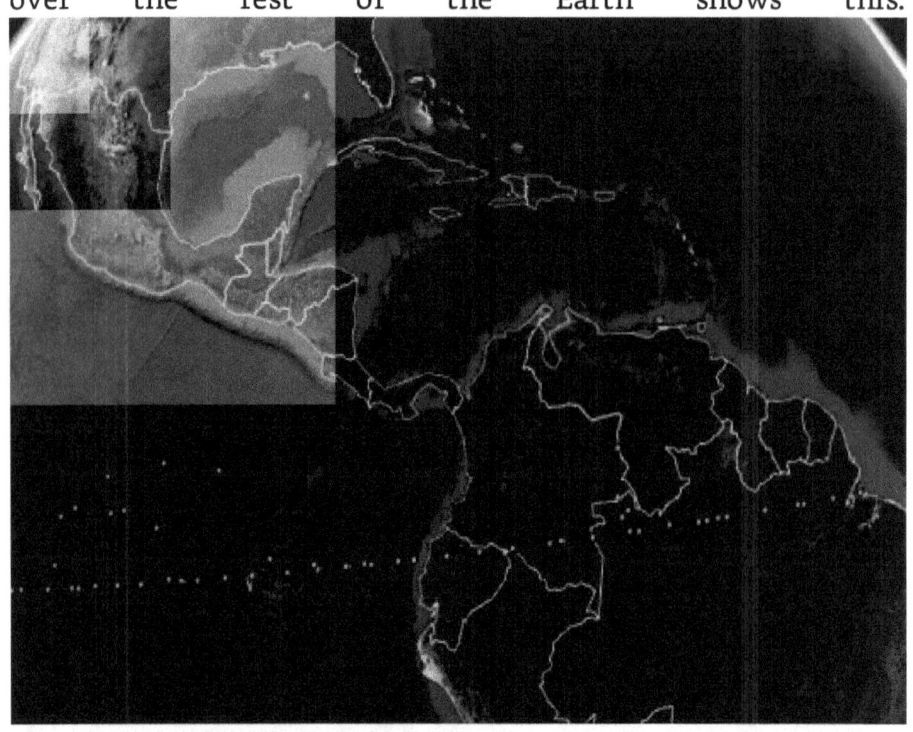

GEO satellites (yellow dots) - Americas
UCS satellite database as of April 2018 projected onto Google Earth

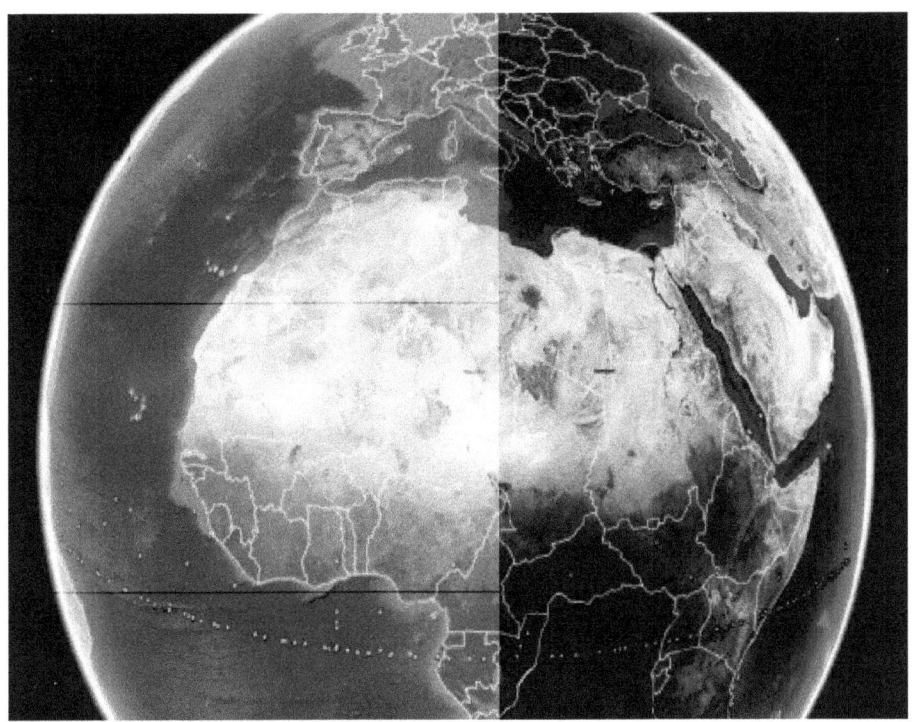

GEO satellites (yellow dots) - Europe/Africa
UCS satellite database as of April 2018 projected onto Google Earth

In summary, while space may look empty, the real estate at GEO is crowded. Since we will place the GEO Station at GEO, and the cable will pass through it, we need to select a GEO slot which provides enough room for the cable to maneuver in all directions. Not just one slot, either - the GEO Station will be large and a window of 20 or more contiguous slots would be ideal.

In turn, the selected spot will determine just where, on Earth, the cable will drop down to. A critical point: the cable can be towed north or south to relocate it, but once in position, we cannot tow it west or east as this would take the Home Station out of its designated slot.

At the moment, the Pacific Ocean is looking good by this criteria, though in a decade or two, GEO is likely to be even more crowded.

8. METEOROLOGICAL CONSIDERATIONS

Ideally, we require a location on Earth with benign weather, where the winds are always calm, there are no hurricanes (typhoons, cyclones), no lightning. Early research indicated the CNT (carbon nanotubes) could be vulnerable to lightning strikes and although we may develop the material or management methodologies to mitigate this, it is better to avoid any issues in the first place.

Amazingly, there are eight locations on Earth where the weather is So, calm and predictable. Sailors, in olden times, were already aware of a number of these locations. The phrase "becalmed" conjures up an image of a seventeenth century galleon, sheets flapping idly, stuck at sea, in hot weather, on a lifeless ocean, as the captain and crew pray for a wind. Satellite data collected over the decade to 2006 by NOAA, used in our previous book, confirmed such locations, back up by an expansion of that data to 2018, which means there are locations that never get storms.

Earlier in the book, we noted a requirement for the Elevator Space Port to lie within 35o of the equator, at most, the closer the better. Now we can overlay this with meteorological data. In the graphic below, the dark grey areas show hurricane storm tracks, the light grey areas show locations of recorded thunderstorms. The white areas have neither.

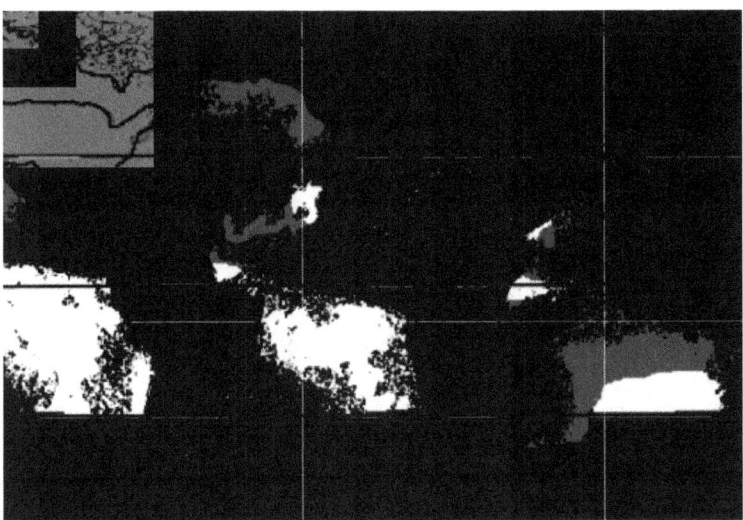

Areas with no storms or lightning (in white)
Source: NOAA data, analysis by the author

In 2018, much more data is available from NOAA and the maps below show all the hurricane (typhoon, cyclone - these are synonyms) tracks in the NOAA database, some going back to 1842. Hurricanes do not cross the Equator - if they start in the northern hemisphere they stay there, as is true in the southern hemisphere. Once hurricanes develop, the Coriolis force gets them spinning, and then, during their life, they move away from the Equator, heading north or south respectively, as residents of Florida and North Carolina can attest, when a hurricane starting life in the West Indies seems to reach the US mainland and maliciously turn north, heading for FL and NC.

Atlantic Ocean

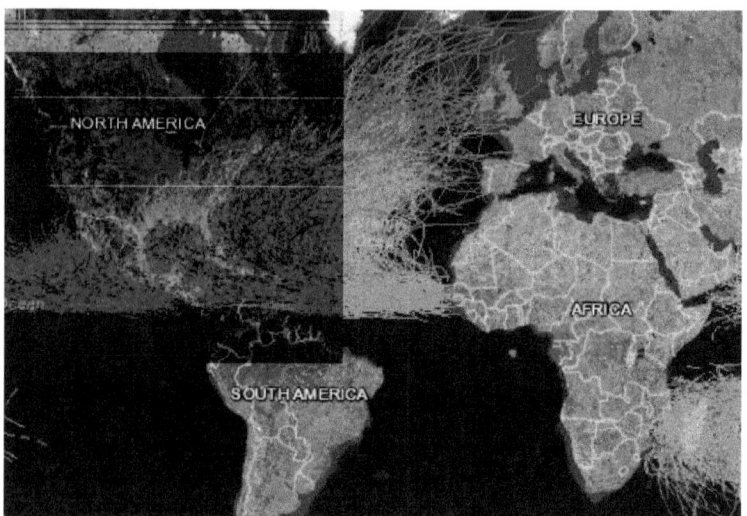

Hurricanes - Atlantic Ocean

Most of the northern Atlantic Ocean is subject to frequent hurricanes, excepting the east side close to the coast of Africa. The southern Atlantic Ocean does get hurricanes, as shown in our early NOAA data, but the data is missing for this segment. However, a large segment of the southern ocean, south of the Equator, is storm-free.

Indian Ocean

The Indian Ocean is mostly a stormy ocean. India, famous for its Monsoon Season, is a storm focus, as are the countries on the latitude of Thailand. South of the Equator, hurricane tracks stretch from Australia to Africa. The apparently clear patch in the middle of the Indian Ocean is the latitude of the Equator. Hurricanes may not happen here, but stormy weather does, including many thunderstorms. There is a peaceful, calm patch, to the south of the Indian Ocean, on the latitudes of Exmouth to Perth, Western Australia, which is a possible Terminal location, though, in terms of latitude, it is on the fringe of acceptable parameters due to the distance from the Equator.

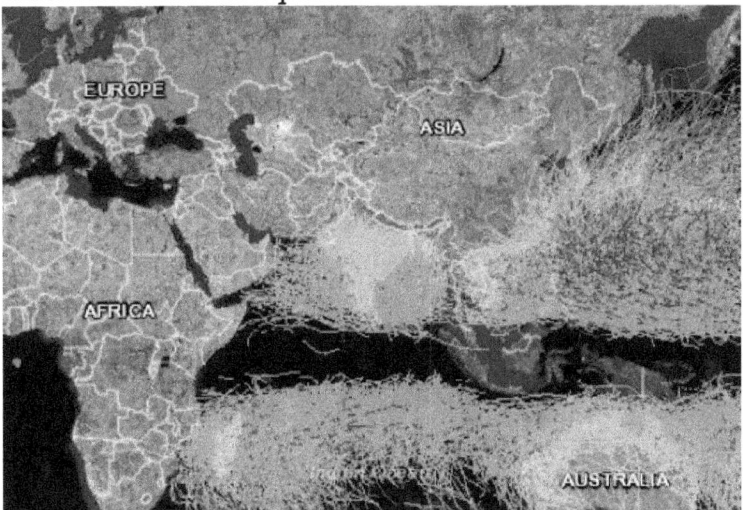

Hurricanes (Cyclones) - Indian Ocean

Pacific Ocean

It is the Pacific Ocean which offers the pick of hurricane-free regions. The line of the Equator stretches from Borneo to Equador, but the portion east of Fiji, extending towards the Galápagos Islands, is mostly storm free. The west coast of America, from California, north to Alaska, faces a hurricane-free zone, which location would be an attractive proposition to the USA, were it not that even San Francisco is 37.7o north, So, that enticing calm patch is rather north for a space elevator.

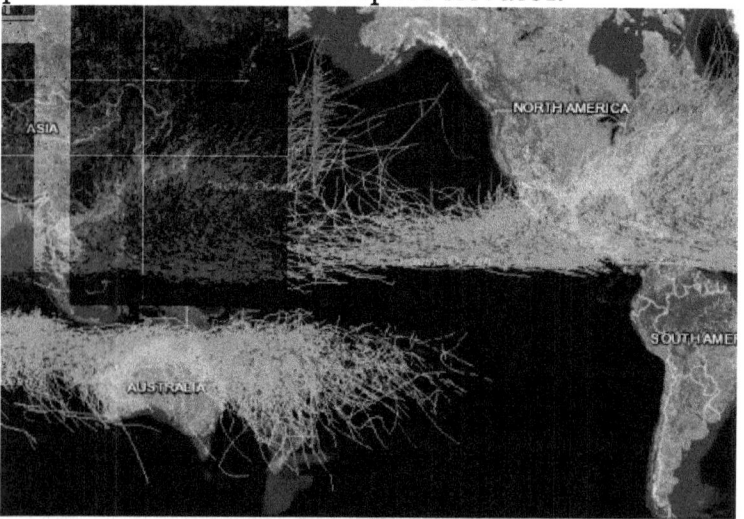

Hurricanes (Cyclones) - Pacific Ocean

It's different for the southern half of the Pacific Ocean, however. From the west coast of South American, from the Equator to the south pole, a massive quarter of the Pacific Ocean offers peaceful, calm, conditions, with no hurricanes or storms. Even the stretch, south of Hawaii, following the Equator west towards Kiribati and the Marshall Islands, offers a peaceful region.

So much for hurricanes, next we overlay thunderstorm activity, based on the lightning flash rate data from NOAA.

Thunderstorms

According to NOAA: 'Typically, more than 2,000 thunderstorms are active throughout the world at a given moment, producing on the order of 100 flashes per second. NASA has two different sensors on satellites measuring flash frequency, the Optical Transient Detector, OTD, and the Lightning Imaging Sensor, LIS. Data from the OTD from 1995 - 2000 and the LIS from 1998 - 2005 has been combined and averaged to create an average annual lightning flash rate map. 11 years of data is included to remove any anomalies that might be present in just one year. The color variations in the map display the average annual number of lightning flashes per square kilometer.

NOAA Lightning flash rates (per year)

The distribution of lightning flashes around the world is uneven. About 70% of all lightning activity occurs in the tropics. The location that receives the greatest number of flashes per year is near the small village of Kifuka in the Democratic Republic of the Congo.

99% of thunderstorms happen over land, or close to land. The map shows, in yellow, orange, red and black, places experiencing ten or more lightning flashes a year, whereas the purple and blue colors, observed mostly over the ocean, indicate no more than one or two flashes a year, per square kilometer.

For the space elevator, in order to avoid any problems from electrical discharges, or interference, with the cable, the preference is for no lightning strikes whatsoever, and the locations in white show zero strikes. Though we caution, there is an element of randomness in thunderstorm activity. Just because a location had no thunderstorms in this eleven year survey doesn't mean it never will have any.

But the white areas are a good starting point for our prospective Elevator Space Port location. Combined with the hurricane data above, it reinforces selections south of the Equator, in the Pacific Ocean, Indian Ocean and the Atlantic Ocean.

For the space elevator, how many lightning strikes per year could we tolerate? There are ways of managing the risk, which will be needed anyway, just in case. We consider a management scheme for two strikes per year can be acceptable, indicatively.

So, deleting locations from the map which receive two or less strikes per year, results in virtually all of the oceans being available for us, excepting inshore areas, particularly on east coasts.

9. OTHER CONSIDERATIONS

Shipping lanes

The Elevator Space Port is a floating platform connected to a cable going skyward, but it presents a potential obstacle to shipping. We really don't want accidents and collisions, So an exclusion zone around the platform will be required, except for shipping servicing the Elevator Space Port.

When we are selecting a particular location, we will need to study local shipping and aviation needs. The good news is, most shipping avoids the middle of the oceans; there aren't any major shipping routes crossing our shortlist of potential locations.

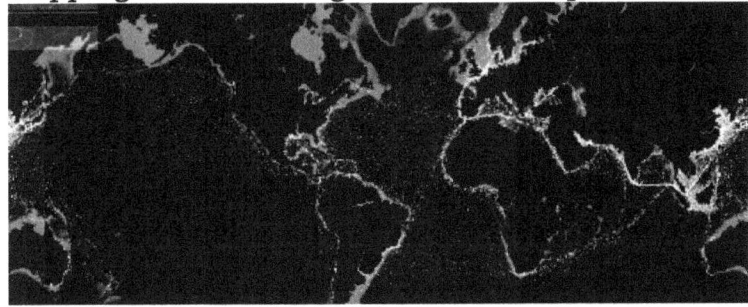

Shipping lanes. Source: www.shipmap.org/

The main shipping lanes are from Asia to Europe, via the Suez Canal or round the Cape, South Africa, and Asia/Japan to USA. It is fortunate, most shipping lanes are north of the Equator, whereas the potential space elevator locations are mostly south of the

Equator, with the exception of shipping around South Africa, and the Europe to Brazil lane.

Honolulu is at latitude 21 degrees north which puts it approximately 2,350 km from the Equator. At a typical sailing speed of 26 knots (48 kph) this gives container ships a journey time of at least 48 hours but that is not unusual in shipping terms. Alternatively, it would be about a 3-hour flight from Hawaii to the Elevator Space Port, if sited at one of the preferred Pacific Ocean locations.

Flight paths

The Elevator Space Port hosts the cable, slicing through the atmosphere into space, So, an exclusion zone for aircraft is required, to avoid any collision risk. Commercial airplanes fly at altitudes of up to 10,000 meters. Exceptions will be any aircraft servicing the Elevator Space Port itself, but flight paths will be set, to avoid the cable.

This graphic from Arcgis shows the main commercial flight paths.

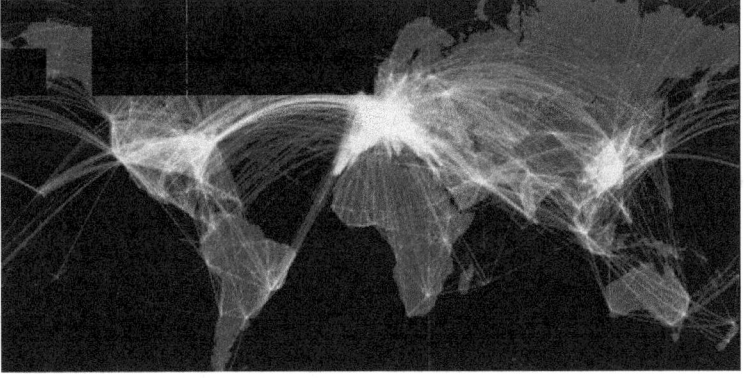

Global commercial flight paths

As for shipping, most flight paths are northern hemisphere paths. If you are not used to visualizing the Equator, it is south of the big bulge in Africa. Traditional maps, based on the Mercator projection, typically have the Equator two-thirds of the way down the map, So, they are heavily skewed to the northern hemisphere.

For our potential space elevator locations, there are few flight issues. The most significant are:

> For any Elevator Space Ports located west of San Francisco, both SF and LA are major international flight destinations.
> Hawaii is a major international flight destination, but it is well north of potential Space Port locations. LA/SF to Australia/New Zealand is a major flight path, though the number of flights is small relative to the major flight paths and flights can be routed

around any Terminals.
> Similarly for the southern Indian Ocean, a small number of flights use the flight path from Perth, Australia to South Africa and Mozambique.
> Any southern Atlantic Ocean terminal locations will take into account the Europe/Brazil flight path, particularly for any Space Port based around Cabo Verde. However, Europe is most likely to express interest in this location, since it already uses French Guiana as a space rocket launch site, So, any European-managed terminal is likely to be well to the west of the main flight path.

Military protection

This is a locality issue too, but an important one. The Elevator Space Port, as the worlds' only cost effective link into space, will have a value beyond price. Unfortunately, in these days of terrorism and national rivalry, it will make a terrorist target.

Protecting a floating target will call for levels of security beyond anything experienced at an airport. The potential for attacks from the air, from ships, or from underwater, likely means a wide exclusion zone will need to be imposed, possibly hundreds of kilometers in radius.

Given its prize value, the size of the defense force will be on a par with America's current best practice, and it will require a large size military support base within close proximity. This may mean locating it near one of the existing American military bases, such as Hawaii, home to the US Pacific Fleet, or Perth Western Australia, home to the US Indian Ocean Fleet, or it may mean creating a new base. This issue rates as a significant location issue.

Considering global control, from a US point of view, this graphic shows where US Military is active:

US military deployment. Source: geocurrents.info

Considering the Pacific and Indian Oceans, global partners of the US include Japan, Australia, Columbia, Peru and, to a lesser extent, Ecuador. The US state of Hawaii is central in the Pacific, of course, and hosts the US Pacific Fleet out of Honolulu. So, support of an American Elevator Space Port in the proposed locations in the Pacific, and in the Indian Ocean, can be included within current deployments.

Other nations will have to contend with American military superiority in these oceans. Mostly, this is not an issue, except for one country - China. When our original book was written a decade ago, China was not seen as a significant military threat, but as of 2018, the US regards the Chinese military as a growing threat.

Given the incredible growth of the Chinese economy, political and economic threats between the USA and China have grown, as testified by two current flash-points: the current account trade "deficit" with China, resulting in the trade sanctions war commenced by President Trump; and the claim by China over the South China Sea and the construction of military bases on what were formally insignificant atolls. China is reaching out globally, expanding its' economic, military and transport footprint.

In the context of this topic, of note has been attempts by China to forge agreements with Pacific island nations, ostensibly for shipping, port and trade operations, but which could be the precursor for Pacific military bases. Questions were raised in April 2018 over a huge port China built for cruise ships in Vanuatu which some suggested could have a military use as well, although this was later rejected by Beijing and Port Vila.

China is situated well to the north of the Equator, having one of the stormiest coastlines. Even the atolls of the South China Sea suffer hurricanes (typhoons) and are 10 to 15 degrees north of the Equator. Equally, Vanuatu is 15 degrees south of the Equator, just outside the ideal weather profile for an Elevator Space Port, but nevertheless a potential location.

Should China decide to base a space elevator in Vanuatu, well, the USA has a problem. One of the best locations for an American space elevator is Kwajalien, an American base in the RMI, Republic of the Marshall Islands. The problem lies in both locations having the same longitude, which means, if two space elevator cables were built, one from Kwajalien, one from Vanuatu, they would compete for the same space location at GEO. The risks of collisions, entanglements, aggressive engagements, would be high and are to be avoided.

Safety and recovery zones

Under this reassuring title, we deal with problem and disaster scenarios: what happens if the cable breaks?

As previously noted, we will have recovery management plans in place. Here we note one fact in particular: due to the rotation of the Earth, if a low section of cable breaks, as it falls it will drift to the west. With the Earth rotating eastwards (i.e. the sun rises in the east, which is the same as the Earth rotating towards the sun) it, in effect, moves out from under the falling elevator cable. This creates the relative impression that the cable falls to the west.

Remember, the cable is light, and has a cross section like a piece of paper. It wouldn't suddenly fall like a brick, but rather would drift down, more like a piece of newspaper floating in the wind. Because of this, it has time to be affected by the rotation of the

Earth and would gradually lie down to the west of the Elevator Space Port and angling towards the Equator. The key plan will be to wind the cable in, to the Elevator Space Port, as fast as it falls, which is quite practical, but in the event of a cable drifting west, a large expanse of open ocean is a plus, in avoiding any entanglement.

(It is different for rockets which are launched at great speed towards the East, and, if they fail, fall back to Earth at a point east of the launch site.)

International airport locations

The Space Elevator can't operate in isolation. People and material will need to access it and an international airport will need to be within easy reach. Fortunately, in these days of ubiquitous air travel, most potential locations will have an airport capable of servicing it.

Even with an air strip, we don't envisage the Elevator Space Port having the capacity to service the largest planes, but providing it can service mid-range jets, access to an international airport should be available within say 2-3 hours flying time.

For example, a location in the Pacific Ocean could be serviced out of Honolulu airport, Hawaii, while a location in the Indian Ocean could be serviced from Perth airport in Western Australia.

Servicing and staffing

Access for servicing and staffing raises the same issues. However, we could assume, in addition to flight access, some servicing via shipping. Heavy goods would go by ship, which means that the Elevator Space Port has port-side harbor facilities. If it was close enough to land, then staff and ground crew could access by ship. We could consider having cruise ships stationed adjacent the Elevator Space Port, offering accommodation for ground crew, as well as entertainment. Such cruise ships would double as "life rafts", giving emergency evacuation capacity.

Environmental issues

Environmental issues are the other key factor. Since it will be a floating platform, we will probably face fewer issues than if we were intending to site it on land. But no doubt there will be issues to address.

If there were islands or atolls nearby, we would need to ensure that there is no adverse environmental impact. For example, the Howland Islands are within an area that is of interest to us for the Elevator Space Port, but the Islands themselves have unique ecologies and are within a US national park. We would take such factors into account in determining a suitable location.

Conversely, the environmental impact should be minor. Overall, it is environmentally positive since it will be offset by a reduction in rocket launches. Sited on a floating platform, the actual cable will be near invisible, and the impact will be less than that of a new airport site. At any rate, we should consider full environment impact reports as part of the process of deciding an anchor site.

Repeating, due to the near elimination of rocket launches, the Space Elevator is environmentally positive.

Meteorological summary

As a generalization, due to the rotation of the Earth, air mass tends to pile up on the east coast of continents. As a first order, the worst weather, hurricanes, cyclones, typhoons, are experienced by the eastern seaboard of the USA, east Africa, China and eastern Asia and the eastern states of Australia. Of course, weather patterns get more complicated than this, but east coast weather is generally more volatile than west coast weather. Contrast storm season in Florida with the beach scene in San Diego!

This results in seas being rougher, choppier, with higher swells, on east coasts. The Atlantic is a stormy and choppy ocean. Early sailors, of the time of Columbus, assumed this was true of all oceans, their experience being limited to the Atlantic and the east coast of the Indian Ocean. So, we understand their surprise on the first crossing of the isthmus of the Americas, when Balboa first climbed the peak of Darien, in modern-day Panama, and saw the western ocean for the first time, as immortalized in the poem by Keats (though he mistakenly attributed the moment to Cortez, not Balboa):

Or like stout Cortez when with eagle eyes
　He star'd at the Pacific--and all his men
Look'd at each other with a wild surmise--
　Silent, upon a peak in Darien.

The ocean that Balboa saw was calm, hence it was named the Pacific Ocean. Pacific it is, for half of its expanse, but had Balboa been viewing it first from China or Japan during a typhoon, he would no doubt have given it a different name.

Comparing with rocket launches at Cape Canaveral

So, what about Cape Canaveral in Florida? Doesn't that operate in a storm area? They launch rockets from there, So, why not the space elevator?

You have to wonder about Cape Canaveral. There are not many areas more prone to tropical storms and lightning! We guess one reason it was put there was because it was the most southerly US mainland location available. But hurricanes rip through Florida every year, So, it's not an ideal location for a space elevator terminal.

Compare the impact of weather on a rocket launch site, versus an Elevator Space Port. For a rocket launch, you only need to achieve a launch window of good weather for a few minutes; yet even then, lots of launches get cancelled at the last minute because of weather.

It was explained well by Popular Science in December 2018, an extract from which is:

'The biggest reason rocket launches get scrubbed is the weather. If adverse weather is enough to cause delays and cancellations for commercial flights, you can certainly bet it's enough to scrub a mission worth hundreds of millions of dollars.'

'And we're not just talking about extreme conditions. When it comes to space launches, engineers on the ground are monitoring a huge slew of different factors. For NASA's standards, there can be no precipitation during launch; winds cannot be faster than about 21 mph from the northeast and about 39 mph from other directions; temperatures below 48 degrees Fahrenheit will force a launch scrub since cold weather can cause ice buildup on the rocket or create problems in some of the equipment (the 1986 Challenger explosion that killed seven crew members occurred because the rocket's rubber O-rings got too cold on the launch pad the night before); and cloud ceilings can't be lower than 6,000 feet in altitude. NASA even has different procedures for launching through cumulus clouds versus cirrus clouds.'

'Lightning is one of the biggest concerns for space launches, which makes sense when you're trying to get a ginormous piece of metal into the air. NASA won't fuel a rocket if there's greater than a 20 percent chance of a lightning strike within a five mile radius of the launch site and won't launch if lightning is observed within 10 miles of the flight path. That radius includes the presence of the cloud that produced the lightning.'

'And it just happens that Central Florida—home to Kennedy Space Center and Cape Canaveral Air Force Station, where much of America's civilian and private spaceflight operations take place—experiences more lightning strikes than any other place in the U.S. An immense amount of work goes into monitoring electrical field activity, and it's all to ensure a little spark won't lead to catastrophe.'

An Elevator, however, is a continuous operation, 24/7, variable weather becomes more of an issue. Those weather conditions are more of a worry for rockets than a space elevator; even so, given the 24/7 operation, the calmer the weather, the better. Those lightning constraints used by NASA would likely be used for the space elevator too. Ideally, we position it somewhere where we never run that risk.

Imagine two or three cable car arrivals in a day, plus all the air and sea traffic that generates, and you get the picture. Five-meter surf looks great on the beach at Hawaii but not when you are on a floating platform!

As would be expected, there is considerable overlap between areas affected by lightning and areas affected by storms.

Early on, Edwards (co-author) postulated an area west of the Galápagos Islands as the preferred location. The Galápagos Islands, part of the Republic of Ecuador, are an archipelago of volcanic islands distributed on the Equator, and north and south of the Equator, in the Pacific Ocean, 900 km west of continental Ecuador.

The Galápagos Islands would be included in the list of potential sites still, although there are two issues. Firstly, being part of Ecuador, it would require political and legal agreement with that country. Secondly, and perhaps the larger issue, the islands form part of a marine reserve, well known to all since the days of Darwin, and any commercial operations in or near the islands may be resisted due to environmental considerations.

But the good news is, travel west from the Galápagos Islands, across the ocean, and there are plenty of alternatives.

10. ELEVATOR SPACE PORT - PREFERRED LOCATIONS

In due course, we envisage a number of space elevators constructed and in operation. It will be So valuable, it would be risky just to construct one - in the event of any failure, we'd have to revert to rockets again, whereas, with at least two space elevators, there is always a backup elevator to use. For the purpose of this book, we assume the first couple of elevators will be owned by the USA, and location constraints are considered with that in mind. However, once it becomes apparent construction of space elevators is feasible, and that the cost of space access plummets to a tenth that of rockets, it can be expected other countries will want to build their own. In fact, some may beat the USA to it. Assuming peaceful relations over the issue are maintained, there is current research into building elevators in China, Japan and others.

So, our initial criteria for an Elevator Space Port location are:

Within 30° to 35° of the Equator - the closer the better
Calm ocean areas, avoiding potential of storms and lightning, and
A location that can be defended and controlled by the USA or a friendly country.

Based on this, the regions of most interest, listed in order of preference, are:

* The Pacific Ocean, south of the Equator
* The Indian Ocean, west of Western Australia
* The Atlantic Ocean, south of the Equator

Drilling down, there are eight locations of particular interest, again listed in order of preference:

* Palmyra Atoll, Jarvis Island, Starbuck Island or similar, west of Galápagos Islands, Republic of Ecuador
* Howland Island and Baker Island, Pacific Ocean
* Kwajalien, a US base in the Republic of the Marshall Islands, Pacific Ocean
* Northern Pacific Ocean, 1,000 km west of Los Angeles or thereabouts.
* Southern Indian Ocean, west of Exmouth or Perth, Australia
* West of French Guiana, Atlantic Ocean
* Ascension Island, southern Atlantic Ocean
* Cabo Verde (Cape Verde), northern Atlantic Ocean

The entire stretch of the equatorial Pacific from the Galápagos Islands, west to Kiribati and Kwajalien, is mostly storm and lightning free. This is the prime choice sector of the Equator with those conditions!

In our original book, a decade ago, we had more or less the same list, with the addition of the Seychelles in the northern Indian Ocean. With the availability of another decades' worth of data since then, we have left this off of the list as being too marginal.

Like real estate, we can assume the best locations, from an operational point of view, will be snapped up first. Whoever launches the first elevator will have the privilege of taking the best location, but once it is successfully deployed, we can imagine the competition over the second-best location.

The southern Pacific Ocean

Since so many of the preferred locations are in the Pacific Ocean, we'll start with a Pacific Ocean primer, since it's important to understand the competing advantages and claims of each location. Much of the Pacific is currently treated as having little economic value but, with the advent of the space elevator and the riches accruing to any Elevator Space Port locations, this will change.

Most people have a limited awareness of the Pacific Ocean, beyond the fact it is the biggest of the oceans, has Hawaii in the middle, and has many idyllic sounding tropical islands.

Tracking the preferred Pacific Ocean locations in more detail, considering sovereignty, politics and logistics, the graphic shows the storm track data over the Pacific, overlaid with the thunderstorm data, and the red box indicates the range of potential Elevator Space Port locations.

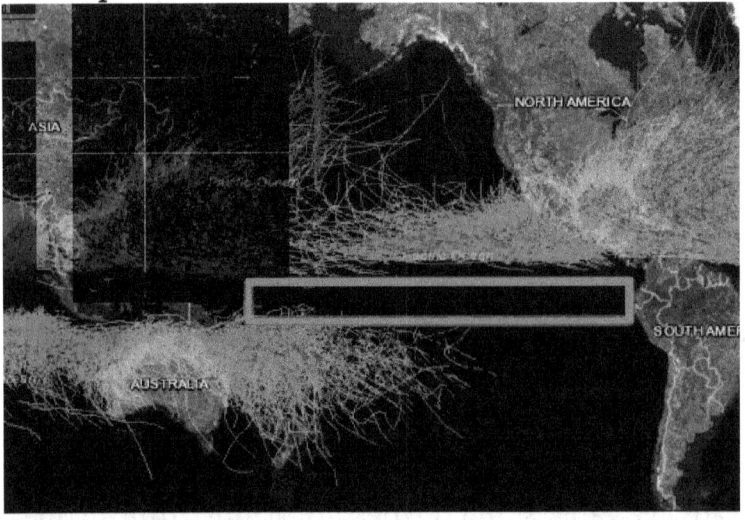

Combined hurricane and thunderstorm data

The area of interest, along the Equator in the Pacific Ocean, is the location of ocean warming in some years, known as El Niño, or the El Niño Southern Oscillation, responsible for diverse, and usually unwelcome, weather conditions at times, such as drought and fires in the USA and Australia.

It is no coincidence we are focused on the patch of Pacific Ocean where the El Niño occurs. Our interest in a location where thunderstorms, lightning, and hurricanes/cyclones/typhoons never happen, where the ocean has benign weather with clear skies, leads us to this area, since the weather conditions that generate El Niño are responsible for the peaceful weather conditions we seek.

East of Asia, however, experiences the opposite conditions. Here, the cold upwelling of water and westerly winds create upward wind cells north of New Zealand which generate storms, cyclones as they are called here. This explains why our preferred section seems to end abruptly in the middle of empty ocean. It is the combination of ocean water temperature variations, and the location of the mid Pacific air updraft, which marks the boundary between benign ocean conditions in the east, and rough stormy weather in the western Pacific.

This boundary is not a fixed point. It moves from year to year. The risk factor, from our point of view, is this: as we approach New Zealand in longitude, the risk of adverse weather increases, so, from a Space Port management point of view, the further east in the Pacific Ocean, the better.

The confluence of weather patterns can be viewed in live NOAA weather satellite images. The image here shows a Pacific Ocean weather composite published by NOAA as at October 2018 (GOES-West RBTop Infrared, Ch. 4 enhanced image):

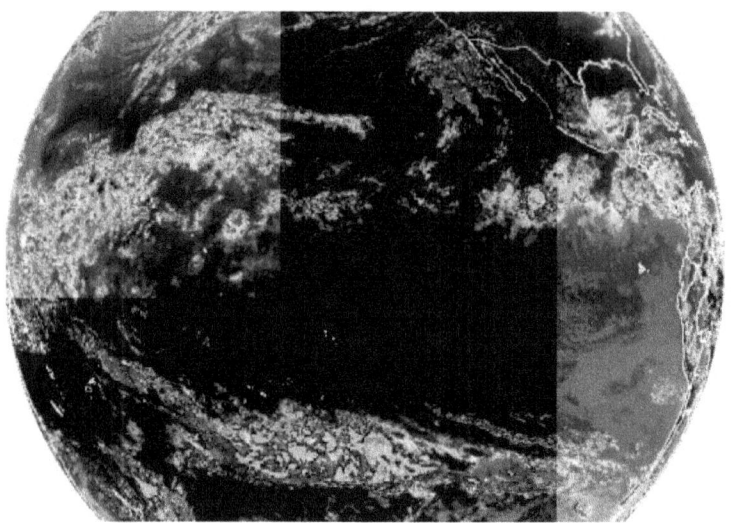
Cloud and storm movements over the Pacific Ocean

In the image, the white speck of the Galápagos Islands can be seen to the right, while the middling black area shows the benign section of the Pacific Ocean which is our target area, along the Equator and into the southern Pacific. The bands of storm clouds, north and south of the Equator, can be seen merging near Asia to the left. Hawaii is in the path of storms. In this image, a cloudy storm area can be seen, west of California while a calm patch of grey is further west in the northern Pacific. This is billed as another potential terminal location, mainly because of its proximity to the USA, but it is still prone to occasional storms passing across it.

International waters

Many of the possible Terminal locations lie in international waters.

There is a patchwork of exclusion zones by country, especially in the eastern Pacific Ocean.

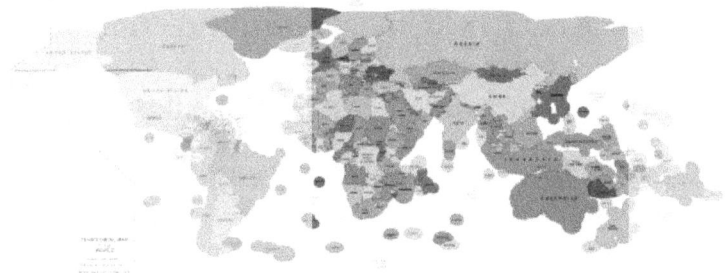

International waters (white) Map created by Rafi Segal and Yonatan Cohen via openDemocracy

'The terms international waters or trans-boundary waters apply where any of the following types of bodies of water (or their drainage basins) transcend international boundaries: oceans, large marine ecosystems, enclosed or semi-enclosed regional seas and estuaries, rivers, lakes, groundwater systems (aquifers), and wetlands.

International waters have no sovereignty, "Terra nullius" as no state controls it. All states have the freedom of: fishing, navigation, overflight, laying cables and pipelines, as well as research.

Oceans, seas, and waters outside national jurisdiction are referred to as the high seas or, in Latin, mare liberum (meaning free sea). The Convention on the High Seas, signed in 1958, which has 63 signatories, defined "high seas" to mean "all parts of the sea that are not included in the territorial sea or in the internal waters of a State" and where "no State may validly purport to subject any part of them to its sovereignty."

The Convention on the High Seas was used as a foundation for the United Nations Convention on the Law of the Sea, signed in 1982, which recognized Exclusive Economic Zones extending

200 nautical miles from the baseline, where coastal States have sovereign rights to the water column and sea floor as well as the natural resources found there.

Ships sailing the high seas are generally under the jurisdiction of the flag state (if there is one); however, when a ship is involved in certain criminal acts, such as piracy, any nation can exercise jurisdiction under the doctrine of universal jurisdiction. International waters can be contrasted with internal waters, territorial waters and exclusive economic zones.' - Wikipedia

'The primary instrument governing the protection of seas is the United Nations Convention on the Law of the Sea (UNCLOS). UNCLOS was adopted at the 1982 UN Conference on the Law of the Sea and came into force, after protracted negotiations, in 1994. It is the "constitution for the seas". The most comprehensive international treaty ever concluded, it establishes rules for all types of use: navigation, fishing, oil and gas extraction, seabed mining, marine conservation and marine scientific research. To date, 165 states and the EU have signed and ratified the Convention.' Source: World Ocean Review

The implications for a floating Elevator Space Port, in international waters, are complex. Such a terminal would be treated as a ship under the jurisdiction of its' flag state. So, a terminal flying the flag of the USA would be sovereign US territory. However, any national state operating a terminal would seek to extend sovereignty beyond the physical infrastructure, at the very least to include an exclusive control zone. The same problem applies to mining and drilling activities in the open ocean, but no definitive rules have been agreed under UNCLOS at the time of writing.

Sovereignty in the eastern Pacific Ocean

Considering the areas of interest in detail, the eastern extent of the Pacific Ocean, to the east of South America, offers the largest contiguous areas of international waters.

Our area of interest commences outside of the EEZ (economic exclusion zone) of the Galápagos Islands, part of the Republic of Ecuador, that is, the international waters extending west of the islands. This patch is remote ocean, and apart from the Galápagos Islands, the closest exclusion zones are Clipperton Island, and Easter Island.

Clipperton Island, to the north, claimed by France. It is 1,000 km south-west of Mexico and uninhabited. It is some 2 sq km in size and largely barren. Should France (or, by proxy, Europe) ever decide to base a Terminal in this part of the Pacific, however, it could become hot property, since it gives France/Europe a legal sovereign base for operations.

Easter Island is a famous island in the southeastern Pacific Ocean, part of Chile, at the south-eastern extreme of the Polynesian Triangle in Oceania. It has a small population. At latitude 27.7o south, it is 3,000 km south of the Equator. That's a bit more than Hawaii is north of the Equator (2,350 km) but if Chile were to consider operating a space elevator, it could prove a useful supply and management location.

For the USA, this stretch of open water on the equator is desirable, but in terms of operating logistics, it is between 4,000 km and 6,000 km from the nearest supply points in southern California or Hawaii. For a country operating globally, this could be managed, but the lack of even a small island to act as a base is a minor negative. It is possible for agreement to be reached with Ecuador to lease one of the Galápagos Islands, which would assist with logistics but be opposed by environmentalists. In any case, there are better choices.

Past relationships between the USA and the Republic of Ecuador have been variable. America had a ten year lease on the Manta Air Base from 1999 but this was not renewed by the then government. In 2014, the Security Cooperation Office based in the US embassy was shut down, forcing some 20 US military personnel to leave the country.

However, as recently as August 2018, RT reported a slight improvement in relations:

'The United States will soon be cooperating with Ecuador in dealing with security issues through a new office. The previous government in Ecuador scaled down military cooperation with the Americans.

The establishment of the new Security Cooperation Office, which was reported by Ecuadorean Defense Minister Oswaldo Jarrin on Thursday, is the latest move by President Lenin Moreno's government to depart form the policies from his leftist predecessor Rafael Correa. Jarrin was careful to stress that the new entity would not violate the country's constitution, which was amended under Correa to ban permanent presence of foreign troops in Ecuador.

"There will be a plane that will allow us to share intelligence. It will stay here for three or four days and then return. So, in no way it means that a new military base is being established," the minister said. He added that due to this arrangement the Ecuadorean Air Force would be handling the cooperation program.'

What the state of relations will be in the 2030's can't be predicted, but better relations would improve the political climate for a space elevator in the international waters west of the Galápagos Islands.

Sovereignty in the western Pacific Ocean

Traveling west across the Pacific, commencing about half-way between the longitudes of Hawaii and the Galápagos Islands, we enter a large area consisting of small Pacific islands of varying sovereignty.

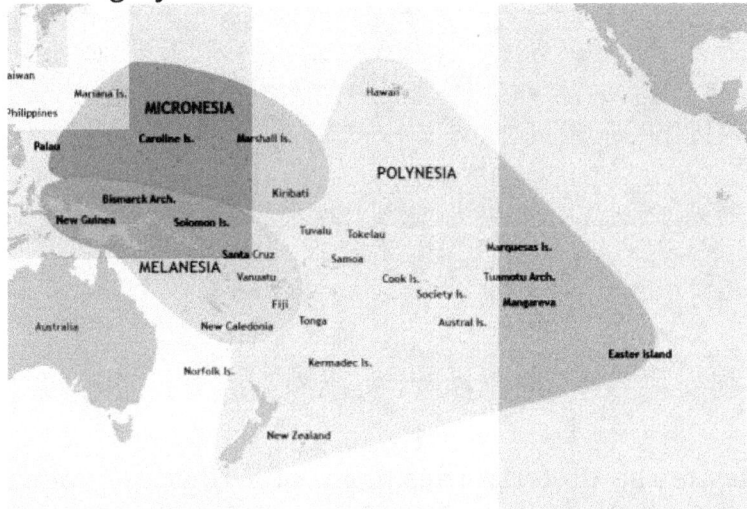

Pacific islands. Source Wikipedia

These are split, broadly, into Polynesia, Micronesia and Melanesia, with different ethnic groups in each. Of interest, Polynesia, excepting New Zealand and Hawaii, represents settlements in the benign ocean areas governed by the easterly winds of El Niño, while Micronesia and Melanesia are settlements leading into the stormier western Pacific. Our interest coincides with Polynesia, ideally, though it will include Kiribati and the Marshall Islands.

Kiribati is going to keep appearing in this survey, because it is a long, straggly line of a nation, stretching along the Equator. Indeed, half of it is called the Line Islands, So, let's identify the location of Kiribati up front.

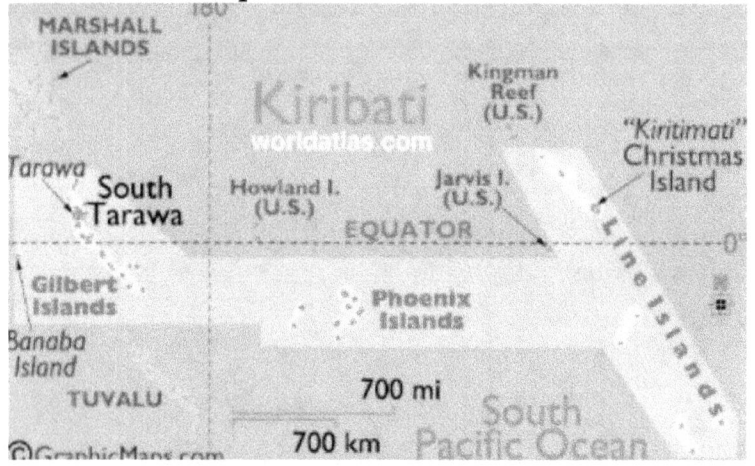

Kiribati, from the Gilbert Islands to the Line Islands (light blue)

Also notable are the US territories north of Kiribati: Howland Island, Jarvis Island, Kingman Reef (with Palmyra Atoll), and Kwajalein in the Marshall Islands.

Many readers will be familiar with key islands, such as Hawaii to the north, Easter Island, a lone sentinel to the east, and Tahiti in the Society Islands. Tahiti and nearby islands are too southerly for our purposes, So, our interest centers on the arc from the Marquesas Islands in the east, through Samoa to the Marshall Islands in the west.

Oceania

Oceania is the name given to the region from Asia/Australia to the mid Pacific, what we describe as the western Pacific Ocean. From our point of view, Oceania contains some of the most desirable locations for the operation of a space port. It has a vast expanse of calm waters and benign weather, combined with an array of small islands which can form operational bases.

However, most of this area is not international waters. Almost all of it is claimed by one or other of the island nations of Oceania, So, issues of ownership come to the fore. It gets complicated. Below we detail each of potential locations, with the most desirable being east of the International Date Line:

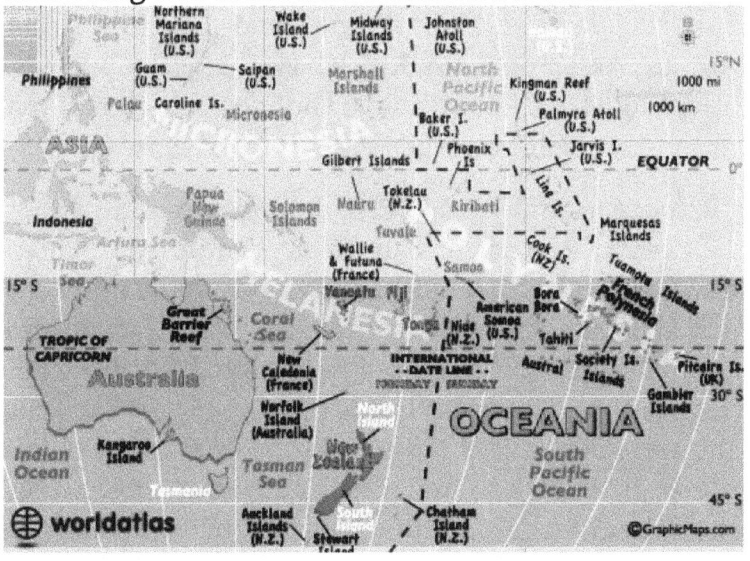

Nations of Oceania. source: Graphicmaps and Worldatlas

* North of the Equator
* Kingman Reef (US)
* Palmyra Atoll (US)
* Howland and Baker Islands (US)
* Republic of the Marshall Islands (Ind.) and Kwajalein (US)

South of, or straddling, the Equator
* Jarvis Island (US)

* Marquesas Islands
* Line Islands (Kiribati)
* Cook Islands (New Zealand)
* Tokelau (New Zealand)
* Phoenix Islands (Kiribati)
* American Samoa (US)
* Samoa
* Tuvalu
* Gilbert and Ellice Islands
* Nauru

This gives us 4 locations north of the Equator, plus 11 straddling, or south of, the Equator.

Some of the most desirable islands, atolls and locations, for a space elevator, are south of Hawaii in what is described as the Pacific Remote Island Area. According to the School of Ocean and Earth Science and Technology at the University of Hawaii:

'The Pacific Remote Island Area (PRIA) includes seven islands located in the Central Pacific that are under the jurisdiction of the United States. Baker, Howland, and Jarvis Islands, Johnston Atoll, Kingman Reef, and Palmyra Atoll, lie between Hawai'i and American Samoa and are administered as National Wildlife Refuges (NWR) by the U.S. Fish and Wildlife Service (USFWS) of the Dept. of the Interior (DOI). Wake Island, which is located between the Northwestern Hawai'ian Islands and Guam, is an unincorporated territory of the U.S. that is administered by the DOI and the U.S. Air Force.'

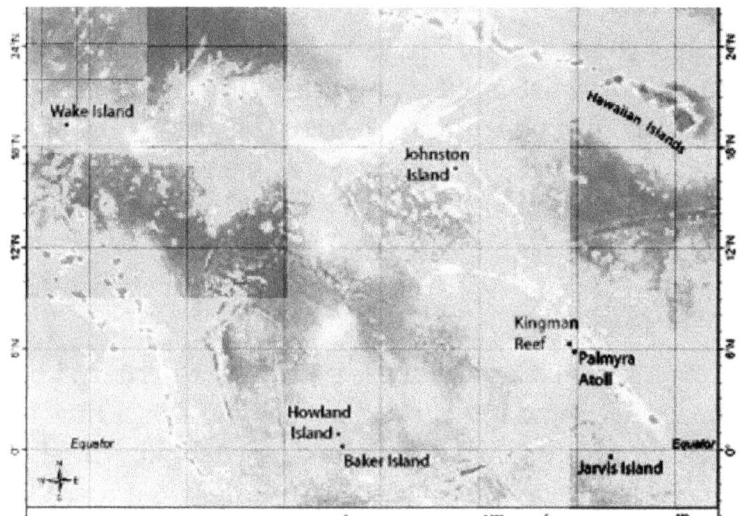

Pacific Remote Island Area. Source: SOEST

(a) Kingman Reef (US)

Kingman Reef, 6°23′N 162°25′W, called Danger Reef in the past, is a largely submerged, uninhabited triangular shaped reef, 9.5 nautical miles (18 kilometers) east-west and 5 nmi (9 km) north-south, located in the North Pacific Ocean, roughly halfway between the Hawai'ian Islands and American Samoa.

It is the northernmost of the Northern Line Islands and lies 36 nautical miles (67 km) northwest of the next closest island (Palmyra Atoll), and 930 nautical miles (1,720 km) south of Honolulu.

The highest point on the reef is less than 1.5 meters above sea level, which is wetted or awash most of the time, making Kingman Reef a maritime hazard. It has no natural resources and supports no economic activity.

Photographer Susan White described Kingman Reef as 'pretty close to the middle of nowhere'.

On this tour we are showing you plenty of candidates for 'nowhere'!

Kingman Reef has the status of an unincorporated territory of the United States, administered from Washington, DC by the US Department of Interior. The atoll is closed to the public. For statistical purposes, Kingman Reef is grouped as part of the United States Minor Outlying Islands. In January 2009, Kingman Reef was designated a marine national monument. (Wikipedia)

(b) Palmyra Atoll (US)

Palmyra Atoll, 5°53'1"N 162°4'42"W, is one of the Northern Line Islands (southeast of Kingman Reef and north of Kiribati Line Islands), located almost due south of the Hawai'ian Islands, roughly one-third of the way between Hawaii and American Samoa. The nearest continent is almost 5,400 kilometers (3,400 miles) to the northeast. The atoll is 12 sq km, and it is located in the equatorial Northern Pacific Ocean. It has one anchorage known as West Lagoon.

Palmyra Atoll Photo

It is administered as an unorganized incorporated territory by the United States federal government. It hosts a variable temporary population of a dozen or two "non-occupants", namely staff and scientists employed by various departments of the US Government and other researchers. The atoll consists of an extensive reef, two shallow lagoons, and some fifty sand and reef-rock islets and bars covered with vegetation. It has an airstrip.

This is an incorporated territory, as opposed to Kingman Reef which is an unincorporated territory. Incorporated territories are integral parts of the United States, as opposed to unincorporated territories being merely possessions. (Wikipedia)

(c) Howland and Baker Islands (US)

Here, we go into more detail about Howland Island and Baker Island, since we consider these to be a prime choice as a base to service an Elevator Space Port. Directly south of Hawaii, almost on the Equator, and US territory, they offer an optimal choice for the USA.

Baker Island, 0°11' N 176°28' W, is just 22 km north of the Equator, an uninhabited atoll about 3,090 km southwest of Honolulu. The island lies almost halfway between Hawaii and Australia. Its nearest neighbor is Howland Island, 68 km to the north-northwest; both have been claimed as territories of the United States since 1857.

The island covers 2.1 sq km. The climate is equatorial, with little rainfall, constant wind, and strong sunshine. The terrain is low-lying and sandy: a coral island surrounded by a narrow fringing reef with a depressed central area devoid of a lagoon with its highest point being 8 meters above sea level. The island now forms the Baker Island National Wildlife Refuge.

It is an unincorporated and unorganized territory of the US which vouches for its defense. It is visited annually by the US Fish and Wildlife Service. For statistical purposes, Baker is grouped with the United States Minor Outlying Islands. Baker Island and Howland Island are alSo,the last pieces of land that experience the New Year.

A cemetery and rubble from earlier settlements are located near the middle of the west coast, where the boat landing area is located. There are no ports or harbors, with anchorage prohibited offshore. The narrow fringing reef surrounding the island can be a maritime hazard, So,there is a day beacon near the old village site. Baker's abandoned World War II runway, 1,665 meters long, is completely covered with vegetation and is unserviceable.

The United States claims an exclusive economic zone of 200 nautical miles (370 km) and territorial sea of 12 nmi (22 km) around Baker Island. (Wikipedia)

Howland Island, 0°48′25 N 176°36′59 W, is an uninhabited coral island located 100 km north of the Equator, about 1,700 nautical miles (3,100 km) southwest of Honolulu.

The island lies almost halfway between Hawaii and Australia and is an unincorporated, unorganized territory of the United States. Geographically, together with Baker Island it forms part of the Phoenix Islands and is grouped as one of the United States Minor Outlying Islands.

It covers 4.50 sq km, with 6.4 km of coastline, which makes it twice as large as Baker Island. The island has an elongated plantain-shape on a north–south axis. There is no lagoon.

Howland Island National Wildlife Refuge is managed by the US Fish and Wildlife Service as an insular area under the US Department of the Interior and is part of the Pacific Remote Islands Marine National Monument.

The atoll has no economic activity. It is perhaps best known as the island Amelia Earhart was searching for but never reached when her airplane disappeared on July 2, 1937, during her planned round-the-world flight. Airstrips constructed to accommodate her planned stopover were subsequently damaged, not maintained and gradually disappeared. There are no harbors or docks. There is a boat landing area along the middle of the sandy beach on the west coast, as well as a crumbling day beacon. The island is visited every two years by the US Fish and Wildlife Service.

Bathymetric data for Howland Island shows a typical atoll formation. Within the proximity of Howland Island, sea depths fall to 500 meters, before dropping off rapidly to 1,000 meters, 2,000 meters and more.

From our point of view, the sea to the west of Howland Island is a potential seabed location for an Elevator Space Port. Our Terminal needs flexibility to move north and south, So, it could be anchored in 500 meters of water, or able to float and move, while being serviced from a base on Howland Island.

The benthic habitat is typical of extinct volcanoes, the land slopes steeply away from the island, but there is a shallower ridge extending to the north-west.

(d) Republic of the Marshall Islands (Ind.) and Kwajalein (US)

The Republic of the Marshall Islands (Marshall Islands) is an island country and a United States associated state, 790 km from the Equator and slightly west of the International Date Line. It is at 7°7′N 171°4′E. By comparison, Hawaii is three times further away from the Equator. It comprises 29 coral atolls, 1,156 individual islands and islets, including the infamous Bikini Atoll which, in addition to giving its name to the iconic piece of swimwear, was the site of the early testing of atomic bombs after WW2.

It achieved independence from the USA in 1979 and entered into a compact of free association in 1986. With a population of over 50,000 people, it is one of the larger Pacific island settlements.

The capital island is Majuro. A concern here is the effect of climate change on sea levels, and rising sea levels are threatening the long term future of Majuro.

Of particular interest is the island of Kwajalein, or "Kwaj" as it gets called.

Kwajalein. Source KRSJV

Leased to the USA, it used to be known as Bucholz Army Airfield in WW2 and is currently utilized by the US for the range testing of missiles in the Pacific Ocean. It is a compete American airbase, offering facilities on a par with US mainland cities, including schools and an airport. Access is restricted to serving US personnel and approved ground staff - other civilians, foreigners and tourists cannot deplane and visit Kwaj.

According to the local web site Kwajnet:

'Kwajalein Island is the largest and southernmost island in the island chain of Kwajalein Atoll, Republic of the Marshall Islands (RMI). Although it is the largest island of the atoll, it is a mere 3 square miles in size. It is commonly referred to by local residents as "Kwaj". Kwajalein is leased to the United States government by the RMI and is a fully functional Army base. The central Pacific location is ideal for strategic defense operations.

USAG-KA's main mission is to support the Ronald Reagan Ballistic Missile Test Site (Reagan Test Site/RTS), which uses radar, optics and telemetry for space surveillance/object detection and ICBM/anti-ICBM missile testing. One of the newer projects, currently under construction, is Space Fence, which will be capable of tracking near-Earth-orbit satellites and space debris as small as 30 inches.'

It has been used by SpaceX for earlier rocket launches.

With the extensive facilities of Kwaj, it would be an obvious choice by the US to include in the shortlist of space elevator locations, since the existing infrastructure can support it.

(e) Jarvis Island (US)

Jarvis Island, 0°22'S 160°01'W, formerly known as Bunker Island, or Bunker's Shoal) is an uninhabited 4.5 sq km coral island, about halfway between Hawaii and the Cook Islands. It is an unincorporated, unorganized territory of the United States, administered by the United States Fish and Wildlife Service of the United States Department of the Interior as part of the National Wildlife Refuge system.

Unlike most coral atolls, the lagoon on Jarvis is wholly dry. While a few offshore anchorage spots are marked on maps, Jarvis island has no ports or harbors, and swift currents are a hazard. There is a boat landing area in the middle of the western shoreline near a crumbling day beacon, and another near the southwest corner of the island. The center of Jarvis island is a dried lagoon where deep guano deposits accumulated, which were mined for about 20 years during the nineteenth century. The island has a tropical desert climate, with high daytime temperatures, constant wind, and strong sun. Nights, however, are quite cool. The ground is mostly sandy and reaches 7 meters at its highest point.

One time, there was an idea for a Jarvis Space Center but nothing came of it.

(f) Marquesas Islands

The Marquesas Islands are a group of volcanic islands in French Polynesia, an overseas collectivity of France in the southern Pacific Ocean. The Marquesas are located at 9.7812° S, 139.0817° W and are 1,000 km south of the Equator.

The Marquesas Islands form one of the five administrative divisions of French Polynesia. The capital of the Marquesas Islands administrative subdivision is the settlement of Taiohae on the island of Nuku Hiva. Over 9,000 people live there.

The island of Fatu Hiva was made famous in a book of the same name by Thor Heyerdahl of Kon Tiki fame.

The Marquesas Islands group is one of the most remote in the world, lying about 1,370 km northeast of Tahiti and about 4,800 km away from the west coast of Mexico, the nearest continental land mass. They fall naturally into two geographical divisions: a northern group and a southern group. There are four airports in the Marquesas, one each on the islands of Nuku Hiva, Ua Pu, Ua Huka, and Hiva 'Oa. With a combined land area of 1,049 sq km, the Marquesas are among the largest island groups of French Polynesia, Nuku Hiva being the second largest island in the entire territory, after Tahiti.

With the exception of Motu One, all the islands of the Marquesas are of volcanic origin. In contrast to the tendency to associate Polynesia with lush tropical vegetation, the Marquesas are remarkably dry islands. Though the islands lie within the tropics, they are the first major break in the prevailing easterly winds that spawn from the extraordinarily dry (from an atmospheric perspective) Humboldt Current. Because of this, the islands are subject to frequent drought conditions, and only those that reach highest into the clouds (generally, 750 m above sea level) have reliable precipitation.

As a French Territory, and being just 1,000 km south of the Equator, this would be an attractive base for a French space elevator. France has been active in space launches for decades, primarily from French Guiana, but for France (and Europe, of which France is part), the Marquesas would be an obvious location to operate from.

(g) Line Islands (Kiribati)

The Line Islands, (known as the Teraina Islands or Equatorial Islands), are a chain of atolls and coral islands, and are located in the central Pacific Ocean, south of the Hawaiian Islands, around 2°S 156.5°W, and (excluding US territories) extending from Teraina, 4.68° north, to Flint island, 11.43° south.

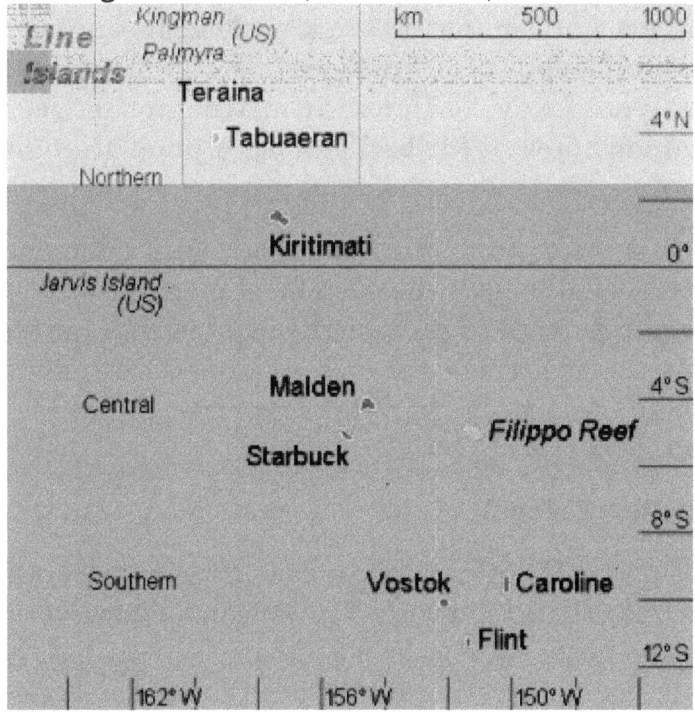

Line Islands map. Source Wikipedia

The 11 islands stretch for 2,350 km in a northwest–southeast direction, making it one of the longest island chains of the world. Eight of the islands form part of Kiribati, while the remaining three (Kingman Reef, Palmyra Island and Jarvis Island) are United States territories. Only Kiritimati and Tabuaeran atolls and Teraina Island have a permanent population.

At the time of writing, Air Kiribati flies between Teraina, Tabuaeran and Kiritamati islands, north of the Equator.

The islands include Starbuck Island at 5.64° south, surely an opportunity for sponsorship of naming rights for a space elevator!

Just north of the Equator lies the main island of Kiritamati, known as Christmas Island, at 1.53° north, 230 km from the Equator. As it is populated and has an air service, it is a potential space elevator location. Kiritamati has the greatest land area of any coral atoll in the world, about 388 sq km and its lagoon is of similar size. The atoll is about 150 km in perimeter, while the lagoon shoreline extends for over 48 km Kiritimati comprises over 70% of the total land area of Kiribati and has a population of about 6,500 people.

Climate change is a major topic for Kiritamati, with concerns that rising sea levels will engulf the island and many futuristic plans have been put forward to protect the main islands. As an independent nation, a space elevator located around Kiritamati would be a valuable asset, generating funds for protection against rising sea levels.

(h) Cook Islands (New Zealand)

The Cook Islands are a self-governing island country in the South Pacific Ocean in free association with New Zealand, due west of French Polynesia and east of American Samoa, while Kiribati is to the north. New Zealand is responsible for the Cook Islands' defenses and foreign affairs, but they are exercised in consultation with the Cook Islands. In recent times, the Cook Islands have adopted an increasingly independent foreign policy.

The Cook Islands' main population centers are on the island of Rarotonga with a population over 10,000 people, where there is an international airport. It comprises 15 islands whose total land area is 240 sq km. Rarotonga, the main island, is located at 21°12'S 159°46'W, so, although it is in an acceptable location So far as weather patterns are concerned, it is of a greater distance from the Equator compared to other locations.

The most northerly atoll of the group is Penrhyn which, at 9°00'20" S 157°58'10" W, is considerably closer to the Equator than Rarotonga. It is an inhabited atoll with a land area of 10 sq km and with a population of about 200 people.

(i) Tokelau (New Zealand)

Tokelau is north of Samoa, at 9.2° S, 171.85° W, south of Jarvis Island and Baker Island.

Tokelau is a dependent territory of New Zealand. It is at about the same latitude as Penryhn Atoll but there is 1,500 km of open ocean between them. It consists of three tropical coral atolls (Atafu, Nukunonu and Fakaofo), with a combined land area of 10 sq km. It lies north of the Samoan Islands, east of Tuvalu, south of the Phoenix Islands, southwest of the more distant Line Islands, and northwest of the Cook Islands.

Swains Island is geographically part of Tokelau but is subject to an ongoing territorial dispute and is currently administered by the United States as part of American Samoa. Tokelau has a population of approximately 1,500 people. It is officially referred to as a nation by both the New Zealand government and the island government.

(j) Phoenix Islands (Kiribati)

By the time I get to Phoenix ... I'll be in Kiribati, again!
Not to be confused with Phoenix Island (singular) in Hainan, China.

The Phoenix Islands or Rawaki are located at 4°30'S 172°0'W, a group of eight atolls and two submerged coral reefs, lying in the central Pacific Ocean east of the Gilbert Islands and west of the Line Islands. They are a part of the Republic of Kiribati. The group is uninhabited.

The United States unincorporated territories of Baker Island and Howland Island are often considered northerly outliers of the group, in the geographical sense, but Howland and Baker are part of the United States Minor Outlying Islands.

The United States previously claimed all the Phoenix Islands under the Guano Islands Act.
The Treaty of Tarawa released all US claims to the Phoenix Islands, excluding Baker and Howland. Kanton (also Canton or Abariringa) Island, is the northernmost and sole inhabited island in the Phoenix group. It is a narrow cable of land 9 sq km.

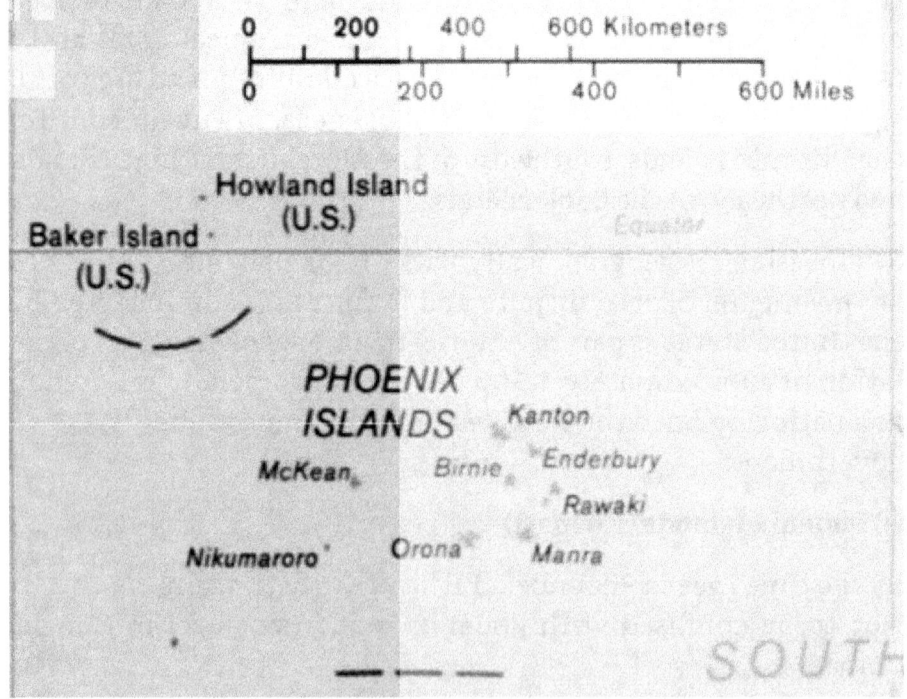

Phoenix Islands map. Source Wikimedia
Note the location of Howland and Baker Islands to the north and on the Equator

Once an important trans-Pacific airport and refueling station, Kanton declined in importance with the introduction of long-range jet aircraft in the late 1950s and was eventually abandoned after serving a brief stint as a U.S. missile-tracking station.

The island still exhibits the remains of the airline and military presence, with 20 people living there. Considering the location as an elevator space port, it shares the advantages of Howland and Baker Islands, just 630 km to the north-west, but whereas they are US territories, Kanton is part of Kiribati.

(k) American Samoa (US)

American Samoa is an unincorporated territory of the United States located in the South Pacific Ocean, southeast of Samoa. It's location is centered around 14.2710° S, 170.1322° W.

It is on the eastern border of the International Date Line, while independent Samoa is west of it. It consists of five main islands and two coral atolls. The largest and most populous island is Tutuila, with the Manu'a islands, Rose Atoll, and Swains Island included in the territory. It is some 500 km south of Tokelau. To the west are the islands of the Wallis and Futuna group. Around 55,000 people live there and they are mostly bilingual in Samoan and English. (Wikipedia)

The seat of government is in Pago Pago. According to amsamoa.net:

'The economy of the Pacific Islands relies heavily on the United States military installations located there. The United States navy, Air Force, marines and Coast Guard all have troops stationed in nearby Guam. Together, the air force and navy operate a research, reconnaissance, and forecasting facility called the Joint Typhoon Warning Center.'

With many naval bases supporting the Pacific fleets, the naval base of Guam is the closest military base to American Samoa territories, though it is over 5,800 km away.

American Samoa is a small territory in comparison to the larger independent Samoa. It is south-south-east of Baker Island. At 1,500 km south of the Equator, the attraction for an elevator space port is the large American shipping presence and infrastructure, but Baker Island offers a location close to the Equator.

(i) Samoa

The independent nation of Samoa is west of American Samoa, across the International Date Line.

The Independent State of Samoa is a country consisting of two main islands Savai'i and Upolu with four smaller islands surrounding the landmasses. The capital city is Apia. The population is about 200,000 people. It shares the location advantages of American Samoa but would offer an independent base for a space elevator for another country.

(j) Tuvalu

Tuvalu is north-west of Samoa, at 7.11° S, 177.65° E, 950 km south of the Equator. It is an independent nation. The Tuvalu islands are small with a population of about 11,000 people.

For a space elevator, it offers a closer proximity to the Equator than does Samoa.

(k) Gilbert and Ellice Islands

Nearby Tuvalu are the Gilbert and Ellice Islands. Once, these were part of the same nation as Tuvalu, now the Gilbert and Ellice Islands are part of Kiribati.

(l) Nauru

Nauru, officially the Republic of Nauru, 0.5228° S, 166.9315° E, is an island country in Micronesia, in the western Pacific. Its nearest neighbor is Banaba Island in Kiribati, 300 km east. It further lies northwest of Tuvalu, north of the Solomon Islands, east-northeast of Papua New Guinea, southeast of the Federated States of Micronesia and south of the Marshall Islands. It has a population of about 11,000 residents in the 21 sq km island.

For years it was mined for phosphate until this was exhausted. It is supported economically by Australia and is currently (2019) the location of what Australia describes as offshore detention centers for refugees, a contentious policy in Australia.

These nations listed, from Samoa to Nauru, are included in the potential list by virtue of their proximity to the Equator. However, they mark the western edge of the benign Pacific weather conditions. Heavy rain and storms can affect these nations at times, So, for space elevator purposes, they are not as attractive as the locations closer to the Equator and further east.

11. BEST SOUTH PACIFIC LOCATIONS

So, for the southern Pacific, how do the locations compare in terms of applicability for a space elevator terminal?

This is our list of locations reviewed:

* North of the Equator
* Kingman Reef (US)
* Palmyra Atoll (US)
* Howland and Baker Islands (US)
* Republic of the Marshall Islands (Ind.) and Kwajalein (US)

South of, or straddling, the Equator
* Jarvis Island (US)
* Marquesas Islands
* Line Islands (Kiribati)
* Cook Islands (New Zealand)
* Tokelau (New Zealand)
* Phoenix Islands (Kiribati)
* American Samoa (US)
* Samoa
* Tuvalu
* Gilbert and Ellice Islands
* Nauru

For the USA, which presumably would prioritize US territories, the choice, running from east to west, is:

* Kingman Reef (US)
* Palmyra Atoll (US)
* Jarvis Island (US)
* Howland and Baker Islands (US)
* Kwajalein (US)
* American Samoa (US)

The US territories comprise small islands and atolls, in the eastern Pacific, then a gap as we jump west, to Howland and Baker Islands, Kwajalein and American Samoa, with Guam (not selected) even further west.

The graphic, with data projected onto Google Earth, shows the locations of GEO satellites (green dots), projected down from GEO, as at October 2018, and compares them with the shortlist of selected locations.

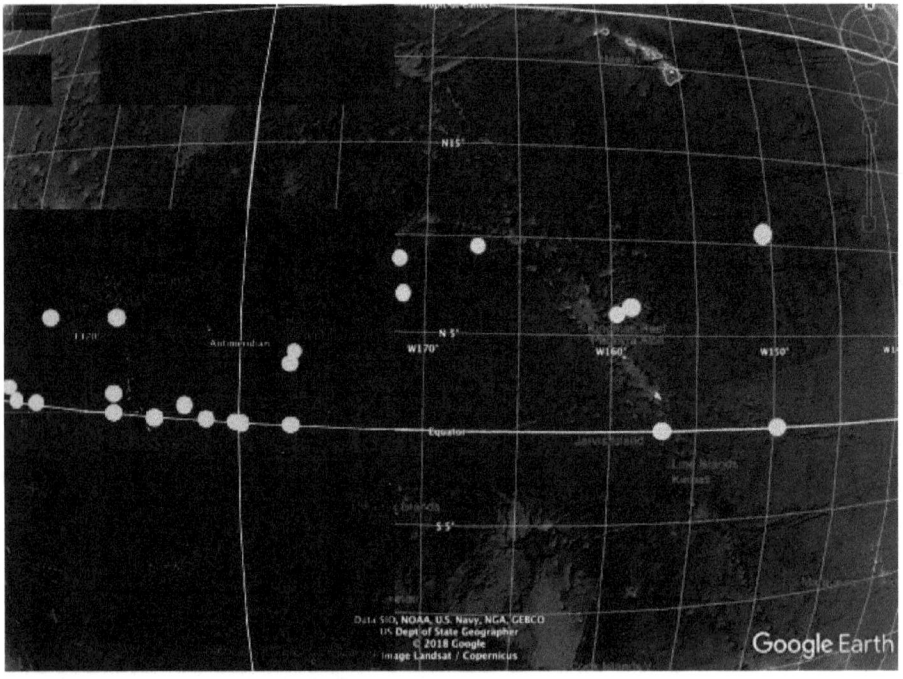

Potential Pacific Ocean locations (red) and GEO satellite footprints (green)

There are very few satellites at GEO, above the longitudes from 140°W to 180°W, that is, almost the whole semi-quadrant east of the International Date Line, which is good news for us, as these slots are available for the intended massive structure of the Home Station at GEO for the space elevator.

Directly south of Hawaii, and east of Jarvis Island, is the projection of the GEO slot occupied by Intelsat 5 (Arabsat 2C, Panamsat-5, PAS-5), a communications satellite launched in 1997. North of that, over Kingman Reef, are two satellites used by the US Air Force, being SBIRS GEO 3 (Space Based Infrared System Geosynchronous 3, USA 273), and SBIRS GEO 4. There are 3 more satellites at the longitude of Howland Island and Baker Island, being DSCS III-F11 (USA), UFO-4 (USA), NSS-5 (Intelsat 803, Netherlands) and NSS-9 (Netherlands), then the number of satellites starts to mount up from Kwajalein west.

This is the situation as at January 2019. By the time we consider deployment of a space elevator, perhaps in the 2030s, some will have come and gone, but the basic premise is likely to be unchanged: these longitudes are least in demand for GEO satellites.

Replacing GEO with LEO satellites

An emerging factor is a trend to launch low earth orbit (LEO) satellites, as opposed to GEO satellites. This may release further slots at GEO, but it comes at the expense of additional objects to avoid at LEO.

The move to LEO has been aided by the maturing of so-called "CubeSat" technology, small satellites the size of a shoe-box. There are several advantages to using CubeSat communications satellites in LEO:

* Cheaper to launch.
* Lower cost per satellite.
* Less distance to transmit signals to/from the satellite, i.e. lower latency.
* Radiation protection from the Van Allen belts means more conventional lower cost electronics can be used.

Set against that, a veritable constellation of satellites has to be launched, to get comprehensive coverage, but even so, they retain a cost advantage. Iridium started this, with a network of 56 LEO telephone communications satellites. More are following. **Telesat** (Canada) reports the deployment of over 117 LEO network of communications satellites at an altitude of 1,000 km, by 2022. **LeoSat** (Florida) has announced intent to launch 108 LEO satellites. The biggest number of intended LEO satellites announced to date, according to **Wikipedia**, include SpaceX Starlink at 4,425 satellites, Boeing at 2,956 satellites, and OneWeb Constellation at 1,980 satellites.

How many such satellites will be in LEO by the 2030's? Currently, LEO satellites and space junk are a nuisance for higher orbital launches, but manageable. If we approach a situation where hundreds of thousands of satellites are in LEO, it will make the job of managing one or more space elevators complex as the need to avoid collisions is magnified.

For space elevator terminal locations, the choice depends on which nation (or group) is operating it. The USA is privileged in having a number of territories in the Pacific. (Incidentally, the movie South Pacific was filmed on the Hawaiian island of Kauai, not one of our Elevator Space Port choices, but the scenery is recognizable on many Pacific islands.)

Best locations

For the USA:

* Best choice looks to be Jarvis Island, on the Equator, directly south of Hawaii, and as we write (November 2018) 30° and sunny. A bare sandy island, uninhabited, it inhabits the most benign spot in the Pacific Ocean. It can be serviced from Hawaii, 2,300 km to the north.
* Kingman Reef and/or Palmyra Atoll. North of Jarvis Island at around 6° N, they share the weather advantages of Jarvis Island, but are not exactly on the Equator, although not far away at 650 km north, which makes them a bit closer to Hawaii, at 1,650 km. Palmyra Atoll is an incorporated territory of the US, as opposed to Kingman Reef which is an unincorporated territory. Incorporated territories are integral parts of the United States, as opposed to unincorporated territories being merely possessions. It means towing a space elevator cable 650 km north once it drops to the Equator, which is manageable. The fact of being an incorporated US territory makes Palmyra Atoll a close second to Jarvis Island.
* Howland Island and Baker Island. These are 1,850 km west of Jarvis Island and are less than 100 km from the Equator. Just east of the International Date Line, they still lie within the benign

weather corridor. They are closer to the Equator than Palmyra Atoll but a longer distance from major service docks: 3,100 km to Honolulu, 1,940 km to Kwajalein.

* Kwajalein, the US island base within the Marshall Islands, is 750 km north of the Equator. It is on the western edge of the benign weather corridor but it does have the advantage of already being a major US military base.

For other countries:

So, what about countries other than the USA? Which ones will be operating in space in the 2030s and have the financial capacity and technical skills to launch a space elevator?

Already on the list are Europe (including France), Russia, Japan and China. By the 2030s, India and perhaps Dubai (UAE) may be included. Other countries with an interest include Australia and New Zealand but are unlikely to have become independent spacefaring nations, though they could operate in partnership with other nations.

French territories in the Pacific include the Marquesas Islands, and the large area of French Polynesia, centered around Tahiti. These are well to the south of the Equator, but east of Hawaii and have potential as bases for a French space elevator, or, a European one.

Japan does not own territories in the Pacific, but it maintains strategic interests. Japan hosts a forum every three years with the leaders of the 14 Pacific Island countries (PICs) which could be a basis for negotiating an operating agreement.

China does not own territories in the Pacific. However, trying to describe Chinese interests in the Pacific is not as simple as it looked a decade ago. In Asia and the Pacific, China is becoming a serious rival to the United States hegemony. If the 20th Century was the American Century, then at some point, it seems likely the 21st Century will become the Chinese Century.

Phillips writes:

'Living as I do in Australia, we Australians are very aware we are living in a time of transition between American power and Chinese power. At the moment, Australia seeks to juggle both powers. It is heavily involved with US military operations, having many US military bases in Australia, and culturally has always been close to the USA. However, in the past decades, trade with China, principally in the supply of raw materials from mining operations, has increased dramatically to the point where Chinese trade is the dominant driver of the economy.'

'Australia depends on the USA for military protection, and China for trade. We know there will come a day when Australia has to choose which partner is more important to it, though when? That is the question. The biggest fear of Australia is of war breaking out over the Chinese claim to the South China Sea. Would we really go to war to defend US access to these waters, or would we acquiesce in Chinas sovereignty claim?'

'Whatever, it seems likely that, by the 2030s, the Chinese presence in Asia and the Pacific will be greater than it is now, perhaps to the point of dominance.'

'So, looking from 2019, the consensus view is of a more powerful, maybe even dominant, China in the Asia-Pacific region. However, it may not turn out that way. Back in the 1980s, there was a time when it seemed like Japan would take over the region. Remember the song "Turning Japanese" by The Vapors, as long ago as 1980? It didn't happen, largely due to Japanese demographics, i.e. an ageing population and the outsourcing of manufacture into Asia. China faces a demographic time bomb, but the outcome may well be different.'

China, as part of its Belt-and-Road initiative, has been active in bringing Pacific islands into its sphere. In May 2018, **Melissa Liberatore** wrote, inter alia:

'There's a sharper focus on China's assertive foreign policy in the region, from island building in the South China Sea to its so-called 'debt-trap diplomacy', including in the Pacific. China has promised much-needed infrastructure projects in countries such as Tonga, Samoa, Vanuatu and the Philippines. Now many counties are indebted to China.'

Samoa is on our list for potential space elevator sites, though being well south of the Equator, not an optimal one, but could be of interest to China.

For other nations, who have to negotiate with independent Pacific countries to operate a space elevator, what are their best choices?

The Line Islands, Phoenix Islands further west, and south of the Equator, and Kiribati in general.

> The Line Islands, with 11 significant islands, stretch across the Equator. Three that we have already referred to (Kingman Reef, Palmyra Island and Jarvis Island) are United States territories, but the rest are part of the nation of Kiribati (including that potential sponsorship deal for Starbuck Island). So, the Line Islands outside of US territories come up on top, as the most attractive location for a space elevator.
> Phoenix Islands, further west, and south of the Equator. These are part of Kiribati.

Either way, we have a feeling Kiribati will become a rich and popular nation once a space elevator is built!

12. OTHER GLOBAL LOCATIONS

Clearly, the South Pacific is a winner, in terms of space elevator locations, but there are other potential locations, in three regions, and we address those next.

> North Pacific
> Indian Ocean
> Atlantic Ocean

13. NORTH PACIFIC

Returning to weather conditions in the Pacific, this image shows the patterns of precipitation (rainfall) for November 8, 2018.

Isobars and rainfall (dark patches) November 8, 2018.
Source weather-forecast.com

A reminder: for a space elevator, we desire calm weather. No thunderstorms - electric charges and lightning can interfere with the CNT cable. No major storms - our cable is in operation 24/7.

November 8, 2018 shows typical Pacific Ocean weather. In the image, the green box marks our desired operations areas, close to the Equator, nice and calm, no rainfall, though it is raining further west, from around Howland Island to PNG.

For reference, the red box shows calm weather in Hawaii, while a low pressure system north of Hawaii is feeding some rainfall northwards.

The yellow box shows a patch of calm weather to the west of mainland USA, and this is another potential space elevator location. The weather here is not as predictable but, by and large, it seems likely we could operate a space elevator here.

This location in the north Pacific Ocean has obvious advantages to the USA. It is offshore from the west coast and could be serviced from, say, Los Angeles or San Diego. The possible Elevator Space Port would be around 1,000 km or more, offshore, So,it would not be visible from the mainland, obviating objections from residents.

Travel times to space, for Americans, would be lower than for other space elevator locations, and it would tie in nicely with existing NASA and private space activities.

The main drawback to this northern Pacific location is the distance from the Equator. At 2,300 km or more, and at latitudes of 32.7° N to 34° N, it is right on the edge of the manageable range. To operate a space elevator, ideally the cable goes straight up into space. As it moves away from the Equator, it starts off at an angle. By the time we get to the latitude of Los Angeles at 34° N, the cable starts off at an angle of 56° to the ground. It adds to the length of the cable, the length of cable in the atmosphere, the distance to be traveled on the cable, and presents operations difficulties.

It may be possible, weather-wise, to compromise and move it further south off of the coast of Mexico, but we reach a point where we say, hey, why not just go to Palmyra Atoll or Jarvis Island?

However, for the USA, the decision may turn out to be political rather than logistical. The same applied to Cape Canaveral, which is situated in about the worst US mainland location possible for a rocket base, in terms of extreme weather conditions, but the politicians of the time were happy with the location.

14. INDIAN OCEAN

Next we consider the Indian Ocean or, more specifically, an area of the Indian Ocean well south of the Equator, about 1,500 km off of the coast of Australia, in a benign weather patch, to be serviced from Western Australia.

In our previous book of a decade ago, Leaving The Planet By Space Elevator, we put this forward as the second choice for a space elevator location. It still ranks up there as a good choice, especially for the Asian nations such as India or China or indeed, Australia itself. However, with the benefit of additional meteorological data to 2018, the equatorial Pacific Ocean locations now look more attractive.

In any case, even if the first space elevator is projected from the Pacific Ocean, the Indian Ocean west of Australia would complement it and make a suitable location for a second space elevator. But there are problems.

Firstly, this area is remote. An Elevator Space Port could be serviced from Perth, Western Australia, which is a large city with international airport and shipping port. An alternative is to service it from Geraldton, 400 km north of Perth, but Geraldton does not (as yet) have an international airport.

Second, the latitudes from the Equator through to about 25° S experience storms, called cyclones in this part of the world, spilling their winds and rain out from their journey across Indonesia or northern Australia.

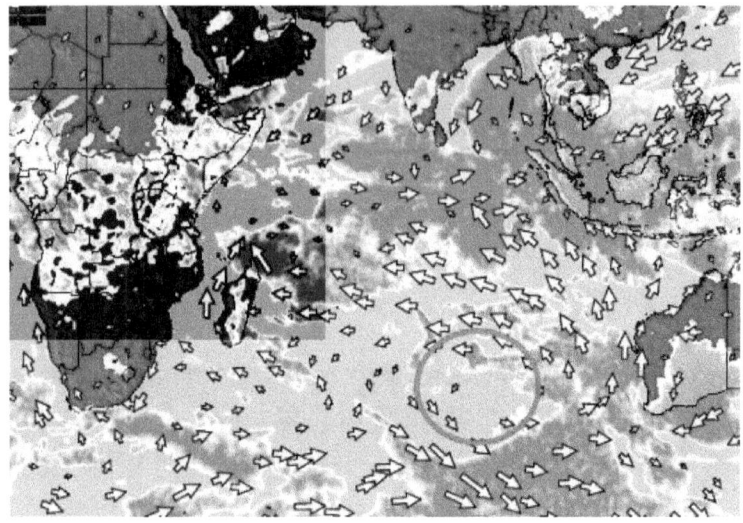

Wind and cloud patterns November 10, 2018.
Source weather-forecast.com

This image, showing the weather patterns as I write, nicely demonstrates the weather patterns of the southern Indian Ocean. A series of high pressure systems stretches from Exmouth in Australia to Madagascar off of Africa, while the area circled red is most commonly at the center of a high pressure system (which rotates anti-clockwise in the southern hemisphere), with clear skies and no storms.

These factors push us down to the latitudes of 20° S to 30° S or thereabouts for a space elevator, which produces the same issues as for the northern Pacific location - it is on the edge of what is manageable for a space elevator operation. This is south of the Tropic of Capricorn at 23° S, though the further north of that, the better.

This base would likely be serviced from Perth, Western Australia, which offers international airport access to global destinations via Sydney, Singapore, Tokyo, etc. Base security would be operated from the Naval Base at Garden Island. It makes it particularly attractive to anyone traveling from China, India or Asia. Perth is located at 32° S, while alternative support bases are available at Geraldton, 28.8° S, or Exmouth, 21.9° S. Exmouth is the location of a secure US communications monitoring base, while Geraldton, in the past decade, has become the operating base for the SKA, the Square Kilometer Array, the new and huge radio telescope array in the Murchison desert, whose chief value lies in the radio silence in this part of the world due to its isolation. The most useful configuration would be to use Perth, Western Australia, as the major support port and city, connecting to Exmouth further north, while Exmouth would be used as the support base for the space elevator terminal.

In this case, the Elevator Space Port would be up to 2,000 km out to sea, and the servicing logistics would be manageable, with sailing times of 20 hours and flight times from Perth of less than three hours - if there is somewhere to land!

Australia is a close ally of the USA still (as of 2019). Perth is the base of the US Indian Ocean fleet, and a modern first-world city similar in size and climate to San Diego, California, except that the surf is better, as Phillips can attest, living there.

The southern Indian Ocean is one of the remotest places in the ocean, in contrast to the northern Indian Ocean. Even further south than our selected area, there is a shipping route from Cape Town, Africa, to Fremantle, Western Australia, which is closer to the Antarctic Ocean than the Indian Ocean. The Indian Ocean is big, bigger than it looks on the standard Mercator map projection used by Americans. The ocean is nearly half the size of the Pacific Ocean.

It is a route popular with adventure sailors and the Golden Globe Race is one such draw for sailors. But its isolation was highlighted in a yachting incident in the race in September 2018. A French research vessel had to rescue injured Indian sailor Abhilash Tomy, after he was left stranded on board his damaged yacht in the remote south-west Indian Ocean, taking him to an island midway between Madagascar and Australia. He was taking part in the around-the-world Golden Globe Race when his 10-metre vessel

Thuriya struck trouble in the south-west Indian Ocean. It took four days for the nearest vessel to reach and rescue him, and he was fortunate the nearest vessel was that close. That is how remote the area is, even to shipping.

Our area of interest for the space elevator lies well to the south. There are no islands out here, nothing but open ocean. The only underwater features are Broken Ridge extending from the Perth Basin, and the N-S NinetyEast Ridge extending along the 90° E longitude.

A drawback is that GEO is relatively crowded over this region, used by China, Asia and India. The image shows our preferred longitude of 90° E, with satellites heavily concentrated to the west, servicing India and Russia, and to the east, servicing Indonesia, Singapore and the rest of Asia.

At around 90° E, as at 2018, there are three Russian satellites, one Japanese satellite and a US Air Force satellite.

On the whole, this location represents an attractive base from which to operate an Elevator Space Port. At 90° E, a space elevator cable would be approximately at right angles to a Pacific Ocean space elevator around the International Date Line at 180° E.

15. SOUTHERN ATLANTIC OCEAN

Another area of potential interest lies in the southern Atlantic Ocean, where there are several potential locations.

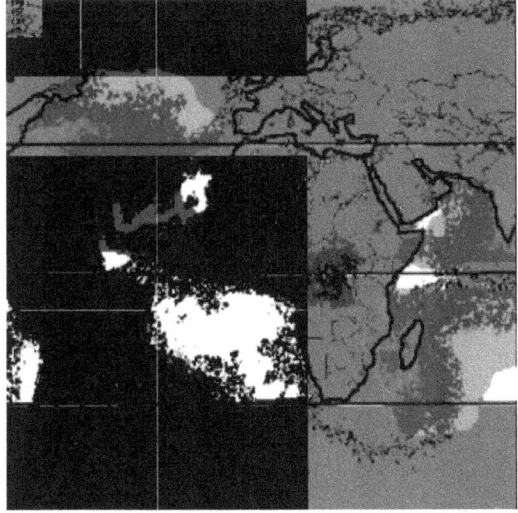

Locations with low storm or lightning activity in white

One is west of French Guiana. Since the French and the European Space Agency (ESA) already operate from French Guiana, this location could well be a priority for Europe.

In addition, much of the South Atlantic looks tempting for a space elevator, though there is the question of geopolitical stability in operating there. Potential agreements with Brazil (South America) or South Africa (Africa) could be considered.

Also, there is a small but interesting patch of calm weather in the north Atlantic, off of the coast of Africa and centered on the Cabo Verde (Cape Verde) Islands.

This image shows the weather patterns in the Atlantic as of November 10, 2018.

A high pressure ridge extends across the mid-Atlantic, with low pressure to the south. Unusually, there is an area of light rainfall east of French Guiana today, in the area of interest, but clear skies to the north and south of this. Also, not typical is the prevailing easterly wind across the Atlantic today. Normally a steady westerly wind blows.

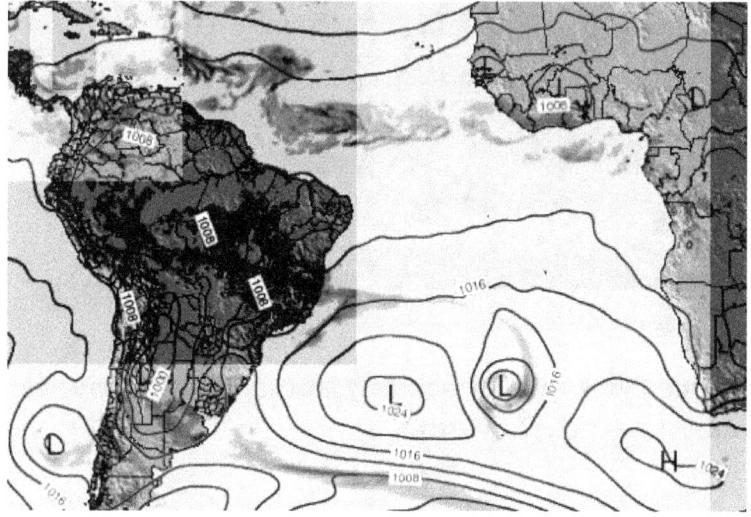

Atlantic Ocean weather, November 10, 2018, showing isobars and precipitation (rain) as dark purple patches
Source weather-forecast.com

It is a reminder that "average" weather conditions are not the same as daily weather conditions. Though, for the space elevator terminal, we aim for locations having the best average conditions for our purposes, we still have to prepare for any and all weather conditions, just like any sailor on the high seas.

French Guiana

The Guiana Space Centre or, in French, Centre Spatial Guyanais (CSG) is a French and European spaceport to the northwest of Kourou in French Guiana, France. It is situated at 5°14''' N 52°45''' W, So, it is close to, but north of, the Equator.

Ariane spaceport, French Guiana.
source: Ariane Space

Operational since 1968, it is particularly suitable as a location for a spaceport as it is near the Equator.

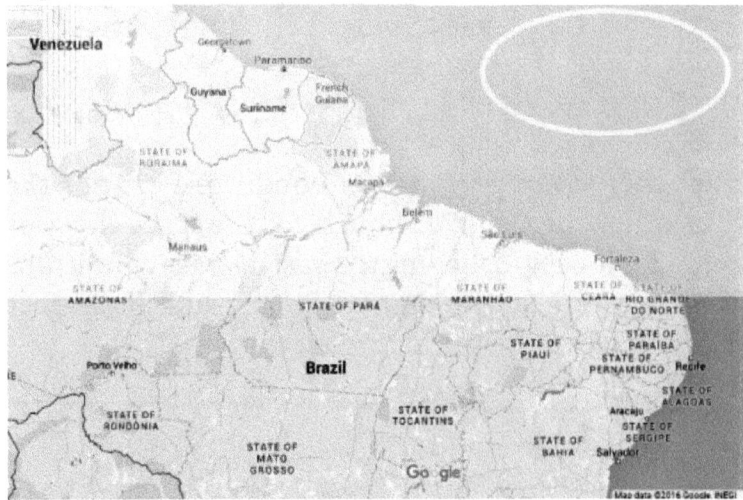

French Guiana, South America,
with a possible space elevator location (yellow)

It is the base of Ariane, and alSo, a number of Russian Soyuz and Vega flights launch from here.

SPACE ELEVATOR EARTH PORTS

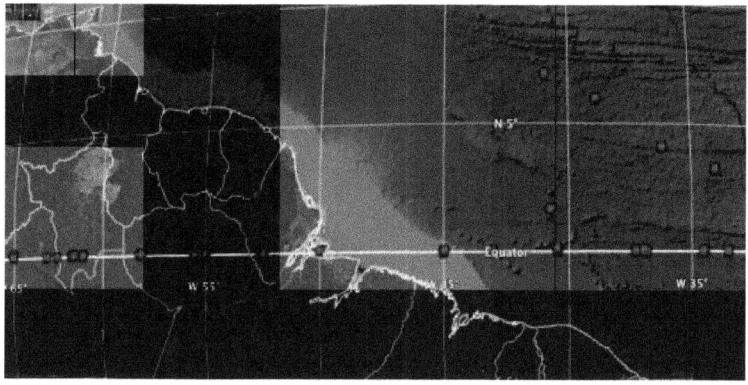

GEO satellite footprints (purple dots) over French Guiana. (Imprinted on Google Earth)

The Atlantic longitudes are popular for GEO satellites but not as crowded as other locations. A potential space elevator terminal could be situated about 1,000 km east of the CSG and managed from there, taking advantage of the existing space port infrastructure.

South Atlantic Ocean

Further afield, there is a vast area of ocean in the southern Atlantic which has potential for space elevator operations owing to the benign weather conditions.

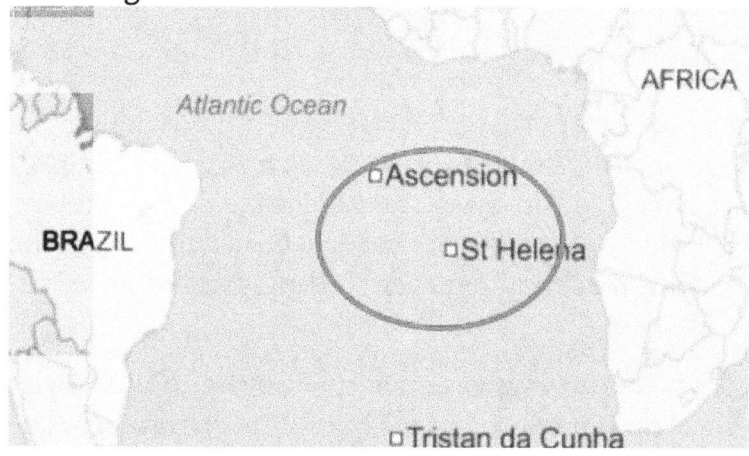

South Atlantic islands and possible locations (red). Source BBC

This is a geologically active area between South America and Africa. It gives potential of an Elevator Space Port around 0° longi-

99

tude, that is, at the Greenwich Meridian, and at longitudes west of it, to about 25° W. At these latitudes, the area faces Brazil in South America, while in Africa it faces Angola, the two Congo's, Gabon and others.

Of particular interest are the two islands in the area of interest: Ascension Island and St Helena. Along with Tristan De Cunha, the three islands are administered as a British Overseas Territory. Assuming they have not become independent by the 2030's, this gives Britain the possibility of operating its own space elevator, and the preferred base would be Ascension Island as the closest to the Equator.

Ascension Island is at 7.95° S, 14.36° W while St Helena is at 24.14° S, 10.03° W. While St Helena is the larger island, Ascension Island is 880 km from the Equator. It has a warm, arid climate, with temperatures around 25° C all year. It already has relevant facilities on the island, being the location of RAF Ascension Island, which is a Royal Air Force station, a European Space Agency rocket tracking station, an Anglo-American signals intelligence facility and the BBC World Service Atlantic Relay Station. The main town is Georgetown with a population of around 800.

An article by the BBC states:

'Officially, nobody is from there. The UK government denies the right of abode, turning Ascension's 800 or so British citizens - some of whom have lived on the island for decades - into temporary visitors. To enter, you must get the written permission of the Queen's representative, known rather chillingly as the Administrator.'

'The airport - whose runway was once the longest in the world, designed to accommodate the Space Shuttle - is operated by the US Air Force, which grants limited access to Britain. Nasa tracked the Apollo Moon landings from Ascension. The European Space Agency monitors rocket launches from here.'

The existing infrastructure and the presence of both the British and the Americans, provide good reasons to consider Ascension Island as a space elevator operating base, with the elevator itself at an Elevator Space Port closer to the Equator.

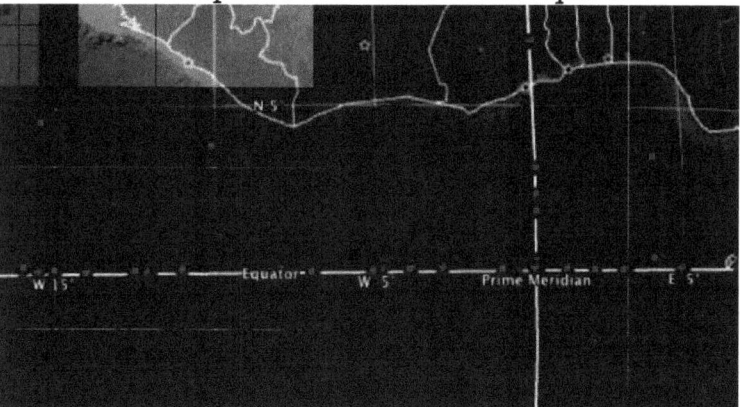

GEO satellite imprint (purple dots) (Imprinted on Google Earth)

A large number of GEO satellites are clustered around the Prime Meridian at 0° longitude, with plenty of others in this region, even around the longitude of Ascension Island near 15° W, So, the actual position of the Elevator Space Port may depend on negotiating a suitable slot.

16. NORTHERN ATLANTIC OCEAN

There is a small patch of benign weather in the north Atlantic, centered around, and to the west of, the Cape Verde (Cabo Verde) Islands. The islands are off of the coast of Africa, which is to the east, the main nations being the desert nations of Senegal and Mauritania.

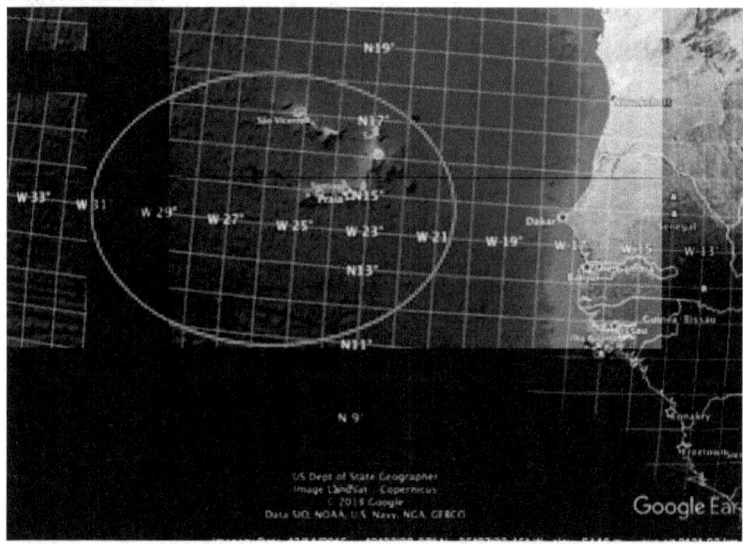

Cape Verde Islands and potential space elevator locations (red)

A cluster of islands, they are located around 23° W, 15° N, and with a potential space elevator location to the west of the islands, they are in a suitable location.

Named the Republic of Cabo Verde, after its Portuguese origins, it is an independent nation, a collection of ten volcanic islands, 570 km west of Africa. Since the early 1990s, Cabo Verde has been a stable representative democracy. Lacking natural resources, its developing economy is mostly service-oriented, with a growing focus on tourism and foreign investment. Its population is around 540,000 people.

The weather is benign here. They are at the same latitude which produces hurricanes of great intensity over the West Indies, Cuba and Florida, but those conditions don't build up until the winds are well to the west, in the Atlantic. This side of the Atlantic experiences calm, dry weather, the same weather which produces the deserts in northern Africa nearby. Cape Verde's climate is milder than that of the African mainland, because the surrounding sea moderates temperatures on the islands and cold Atlantic currents produce an arid atmosphere. It is part of the Sahelian arid belt. Because of their proximity to the Sahara, most of the islands are dry.

Western Hemisphere-bound hurricanes often have their early beginnings near the Cape Verde Islands. These are referred to as Cape Verde-type hurricanes. These hurricanes can become very intense as they cross warm Atlantic waters away from Cabo Verde. The islands themselves have only been struck by hurricanes twice in recorded history (since 1851): once in 1892, and again in 2015 by Hurricane Fred, the easternmost hurricane ever to form in the Atlantic.

This is of interest to us for space elevator purposes. In meteorological terms the location is more unstable than other preferences, but stable enough for our purposes.

Further, according to Wikipedia, 'Cape Verde's strategic location at the crossroads of mid-Atlantic air and sea lanes has been enhanced by significant improvements at Mindelo's harbor (Porto Grande) and at Sal's and Praia's international airports. A new international airport was opened in Boa Vista in December 2007, and on the island of São Vicente, the newest international airport (Cesária Évora Airport) in Cape Verde, was opened in late 2009. Ship repair facilities at Mindelo were opened in 1983. The major ports are Mindelo and Praia, but all other islands have smaller port facilities. In addition to the international airport on Sal, airports have been built on all of the inhabited islands. All but the airports on Brava and Santo Antão enjoy scheduled air service.'

Cabo Verde is not a member of the EU (as at 2018) but is a beneficiary of the EU's regional cooperation programme with Portuguese-speaking African countries (PALOP) countries: Angola, Guinea Bissau, Cape Verde and São Tomé and Príncipe.

At around 23° W, there are a large number of GEO satellites, but many slots are available.

The location, relatively close to the EU, would make these islands attractive for a space elevator. A terminal would have to be towed to around 15° N but this is manageable and better than some of the potential locations on the list.

As an independent nation, the Republic of Cabo Verde would be able to leverage its position for economic gain and perhaps even EU membership.

17. BEST LOCATIONS

So, here we have our list of preferred space elevator locations: four in the South Pacific, three in the Atlantic Ocean, one in the Indian Ocean west of Australia.

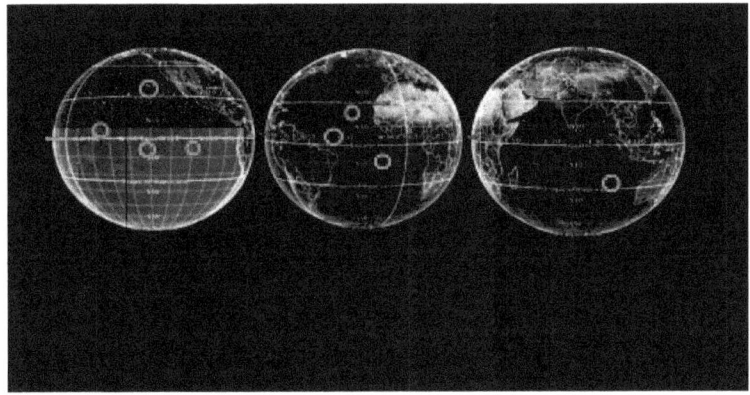

Preferred space elevator locations (red circles imposed on Google Earth)

This gives us a total of eight good locations for an Elevator Space Port, and a number of possibles besides. Which one will be used first? The choice will probably turn out to be a matter of politics and sovereignty.

Which country will be building the first space elevator? The USA, Japan, China, Russia, India, UK, other?

That will influence the attractiveness of any particular location. For the USA, one of the Pacific Ocean locations, maybe Howland Island, Jarvis Island. For France and/or the EU, a base in French Guiana looks an obvious choice. The UK would have Ascension Island, while Japan or China might court Australia for an Indian

Ocean base. However, once a first elevator is constructed and shown to be operational, it won't be long before a second, and more, are constructed, since all of the listed countries will see the advantage of possessing their own elevator.

For some island countries, presently ignored or overlooked for their economy or for the problems of rising sea levels, the space elevator will prove to be a massive financial bonus as they turn into desirable locations.

Once we have two or more space elevators constructed, we will have made a lasting connection to space travel. Never again will we need expensive heavy rockets to get out of the Earth gravity well.

We will be on our way to regular, affordable, space travel, and by the end of the 21st Century we expect space travel to be almost as routine as air travel is today.

SPACE ELEVATOR EARTH PORTS

Book 4 of the Space Elevator 2020 series

Published 2020

Linda J. Phillips

The Space Elevator 2020 series
Book Four:
Space Elevator Earth Ports

Copyright © 2020 Linda J. Phillips
Linda J. Phillips has asserted her right to be identified as the author of this work.

All rights reserved. This book or any portion thereof may not be reproduced or used in any manner whatsoever without the express written permission of the publisher except for the use of brief quotations in a book review.

Photographs, images or artwork are used courtesy of the copyright owners referred to in the source links.

Publisher contact: info@21stcentury.space
FaceBook www.facebook.com/lindyjaniceAuthor
Twitter @_lindaphillips
Amazon author page https://www.amazon.com/author/lindajanicephillips
Web 21stcentury.space

Web links utilized in this publication were correct at the time of writing, but they can change over time.
Published by Linda Phillips

www.ingramcontent.com/pod-product-compliance
Lightning Source LLC
Chambersburg PA
CBHW070657220526
45466CB00001B/470